Technologies of Landscape

Technologies of Landscape

From Reaping to Recycling

Edited by David E. Nye

University of Massachusetts Press Amherst

Printed in the United States of America
LC 99-16762
ISBN 1-55849-228-3 (cloth); 229-1 (paper)
Designed by Dennis Anderson
Set in Adobe Garamond by Graphic Composition, Inc.
Printed and bound by Sheridan Books, Inc.

Library of Congress Cataloging-in-Publication Data

Technologies of landscape : from reaping to recycling / edited by
David E. Nye
p. cm.
"Emerged from a conference 'Landscape and technology,' held at
Odense University in Denmark in January 1997"—P.
Includes bibliographical references and index.
ISBN 1-55849-228-3. — ISBN 1-55849-229-1 (pbk.)
1. Landscape assessment. 2. Nature—Effect of human beings on.
I. Nye, David E., 1946– .
GF90.T43 1999
303.48′3—dc21 99-16762
CIP

British Library Cataloguing in Publication data are available.

This book is published with the support and cooperation of the Danish Humanities Research Council and the University of Massachusetts Boston.

Contents

Acknowledgments ix

INTRODUCTION

Technologies of Landscape 3
David E. Nye

EPHEMERAL LANDSCAPES

1 Agricultural Technology and the Ephemeral Landscape 21
Paul Brassley

2 Journey into Space:
Interpretations of Landscape in Contemporary Art 40
James Dickinson

INVENTING LANDSCAPES

3 Abandoning Paradise:
The Western Pictorial Paradigm Shift around 1420 69
Jacob Wamberg

4 The Employment of the Word: Writing, Topography, and Colonial Landscapes 87
TADEUSZ RACHWAL

5 Remaking a "Natural Menace": Engineering the Colorado River 97
DAVID E. NYE

RESISTING RURAL MODERNITY

6 Begrudging Aesthetics for a New South: The Farm Security Administration Photographic Project and Southern Modernization, 1935 to 1943 119
STUART KIDD

7 Making and Meaning in the English Countryside 136
CHRISTOPHER BAILEY

NARRATING POLLUTION

8 Public Perceptions of Smoke Pollution in Victorian Manchester 161
STEPHEN MOSLEY

9 Narrating the Toxic Landscape in "Cancer Alley," Louisiana 187
BARBARA ALLEN

LANDSCAPE AS PATHWAY

10 Benton MacKaye's Appalachian Trail: Imagining and Engineering a Landscape 20[illegible]
MARK LUCCARELLI

11 "The Landscape's Crown": Landscape, Perceptions, and Modernizing Effects of the German Autobahn System, 1934 to 1941 21[illegible]
THOMAS ZELLER

TOURING LANDSCAPES

12 The Improvement of Arthur Young: Agricultural Technology and the Production of Landscape in Eighteenth-Century England 241

Stephen Bending

13 The Road to Industrial Heterotopia: Landscape, Technology, and George Orwell's Travelogue *The Road to Wigan Pier* 254

Pia Maria Ahlbäck

14 Recycled Landscapes: Mining's Legacies in the Mesabi Iron Range 267

Peter Goin and Elizabeth Raymond

Notes on Contributors 285

Index 287

Acknowledgments

This book emerged from a conference, "Landscape and Technology," held at Odense University in Denmark in January 1997. That event would not have been possible without financial support from the President of Odense University, Henrik Tvarnø, and from the Center for Man and Nature, a five-year research project funded by the Danish Humanities Research Council from 1992 until June 1997. Svend Erik Larsen, then the center's leader, was supportive throughout the planning stages of the conference, and the center staff, particularly Henrik Juel, made sure the event ran smoothly.

There is a long way from a conference with thirty papers to a book with half that many. While the selection of submitted articles and the content of the editing process remained in my hands, the details of preparation were made easier by the assistance of two doctoral students who began their studies shortly after the conference, Ole Bech Petersen and Birgitte Nielsen. Charlotte Granly helped with both the conference and the publication. Two anonymous readers were extremely helpful in putting the papers into perspective and suggesting revisions. My editor, Paul Wright, offered not only a contract but congenial advice. After the University of Massachusetts Press agreed to publish the volume, the Danish Humanities Research Council graciously granted a subvention to assist with the reproduction of the photographs. While all of these people and institutions made the task easier, the final responsibility for any errors remains, of course, mine.

INTRODUCTION

Technologies of Landscape

DAVID E. NYE

As its title suggests, this volume moves from the preindustrial world, when the dominant technologies were agricultural, to the high technology of the present. The focus is on the last 150 years, particularly since 1930 in Britain and the United States. The essays should be read with J. B. Jackson's definition of landscape in mind: "A composition of man-made or man-modified spaces to serve as infrastructure or background for our collective existence."[1] Landscape is thus defined not as natural, but as cultural. It is not static, but part of an evolving set of relationships. Landscapes are part of the infrastructure of existence, and they are inseparable from the technologies that people have used to shape land and their vision.

Rather than treat landscape and technology as opposites, the authors of these essays see a subtle interweaving. Human beings have repeatedly shaped the land to new uses and pleasures, and what appears to be natural to one generation often is the end result of a previous intervention. Today's forested hillside in New England with its brilliant autumn hues was cleared pasture in 1840, and the species of trees and plants we now find there are a mixture of the indigenous and the imported. Some of the apparently wild moors beloved of hikers in Britain were once thickly forested, as was the now mostly open, gently undulating countryside of Denmark. Almost anywhere the historian looks the appearance of the land is the result of a subtle interplay of geological forces, agriculture, industry, and leisure activities. The medium of the air itself is affected, carrying traces of smoke, microscopic particles such as pollen and carbon monoxide, and the dust raised by our journeys. Even at American national parks and supposedly untouched wilderness areas the intermingling of culture and nature continues, whether in the form of acid rain; the migration of plant species; or the creation of maps, guides, trails, lookouts, and other means of framing and representation. None of the authors in this volume thinks in terms of such opposed pairs as wilderness/civilization or nature/culture, for they have found that such dichotomies obscure rather than illuminate the cultural processes that produce landscapes.

Rather than espousing nostalgia for an impossible past when landscapes that human beings produced were somehow more "natural," these essayists take the commonsense position that people invariably and inevitably modify their environment. This does not mean one should approve of all interventions, but it does mean that the idea of wilderness is a fiction, which carries with it considerable useless intellectual baggage. Perhaps most obviously, it implies that Native Americans or Laplanders or Aborigines and other first peoples either were never there or left no trace on the land. Furthermore, to posit wilderness as a real category in nature immediately suggests narratives of preservation, which pit environmentalists against industrial forces seeking to desecrate the earth. While there is some truth to this story in individual instances of mining and logging, this dramatic story is not a useful narrative for understanding landscape, which is not a natural phenomenon, but a cultural creation. As Simon Schama has argued, "Landscapes are culture before they are nature; constructs of the imagination projected onto wood, water, and rock."[2] Not all peoples project the same visions, of course. Germans, the English, and Native Americans do not see the same things when they picture to themselves living in the woods or climbing in the mountains. Schama's *Landscape and Memory* underscores this point with a rich tapestry of examples from history, painting, and literature.

Whereas Schama's starting point is an almost forgotten Jewish burial ground in a corner of the Baltic, William Cronon and his collaborators reached a similar conclusion, in *Uncommon Ground,* by focusing on American sources. As the book's subtitle states, they were "rethinking the human place in nature" and rather quickly realized that "the way we describe and understand the world is so entangled with our own values and assumptions that the two can never be fully separated."[3] The present volume seeks to contribute to this discussion by focusing on the relationship between landscape and technology.

I have found that when these two terms are mentioned together, what comes to many people's minds are the extremes of mechanization, such as the cityscape of Fritz Lang's *Metropolis,* Charles Sheeler's austere paintings of factories, or the futuristic Los Angeles of *Bladerunner.* These alien topographies represent only a tiny fraction of the less obvious but far more pervasive and persistent reconstitution of the spaces we live in. Indeed, I find it impossible to conceive of a landscape that is not the result of human contact with the land. The "natural" scenes that we valorize in paintings, admire when traveling, or seek to preserve from development are in no case untouched. But this does not mean that purity has been despoiled. Quite the contrary. What Western society has traditionally valued in landscape has been the marks of human labor on the soil, visible in cultivation, hedgerows, ornamental plantings, and other activities of *landscaping.* In the English language, *landscape* is a verb as well as a noun, referring to an active process in which human beings do not merely intervene, but improve a site so that it becomes a more useful or pleasing prospect. For example, most American homeowners now feel it necessary to maintain a green lawn, and although this cultural choice is little more than a century old in the United States, the vision of a house surrounded by neatly clipped grass has become naturalized.[4] Yet the lawn is a technological construction. The ground itself has been graded and leveled. The seed is usually a mixture of several kinds of grasses carefully bred and selected. Weeds, that is, the plants that have been socially defined as undesirable, have been eliminated, often by spreading chemicals over the lawn, which usually is fertilized as well. To keep the lawn looking well, the homeowner employs mowers, rakes, and various devices to trim edges and corners. And if it does not rain enough, the homeowner will set up sprinklers. The entire artificiality of a lawn is especially evident in desert communities such as Salt Lake City, where no grass would survive without irrigation, but few lawns anywhere would survive in anything like their present form without constant technological assistance. In short, the lawn is neither natural nor static but the result of active processes of intervention.

The subtitle of the present volume—"from reaping to recycling"—likewise suggests two active processes, one agricultural and the other postindustrial, delimiting the chronology of our discussion. The first landscape paintings, which Jacob Wamberg notes did not emerge until the early fifteenth century, depict agricultural landscapes. Reaping is a seasonal activity, and its depiction immediately suggests the round of the seasons, the rotation of crops from year to year, and a long-term relationship between farmers and the land. Farmland represents an ideal and unchanging landscape only to people who do not live within it. Those who observe it regularly are struck not by constancy, but by change. These may be great transformations, as when Arthur Young in the eighteenth century describes the estates being created by scientific gentleman farmers who were enclosing land and recreating rural life or when the photographers of the New Deal depicted the industrialization and electrification of the American South. But just as important are the transformations in the countryside caused by the introduction of new crops or new harvesting techniques. As Paul Brassley points out, haystacks and bales of hay have disappeared, and some fields previously green are bright yellow with oilseed rape or the Mediterranean blue of linseed. People who are not farmers, particularly in Britain but also in the United States, tend to view such visual transformations as undesirable. City people and suburbanites often wish to keep the country a static repository of tradition, Christopher Bailey reminds us, even if it is against the desires of the rural residents themselves.

When people first began to value rural landscape as an aesthetic and to produce images of it, however, it was not because it embodied history and stood in opposition to urban life. Rather, landscape painting represented a break with visualizing eternal life. The previously dominant tradition of image making during the late medieval period had focused on the spiritual, not the secular, and from this point of view, peasant work in the fields simply was not a subject. When landscape appears as an aesthetic form, it literally registers "a Fall," as Wamberg argues. "Instead of the mountain slopes with their primitive occupations, we are led down and forward toward civilization's flat plains, on which agricultural fields, hedges, roads, and bridges" engage the eye. At its inception, then, to depict the landscape was to show not stasis but process, not the permanent but the ephemeral. Only centuries later, after the industrialization of agriculture, did some people seek to preserve ancient crafts and agrarian tradition.

The second term in the title, "recycling," suggests further complexities. In one sense, of course, recycling is good environmental practice.

But there are other associations with this word. In all parts of the world today, technologies are being used to rebuild, reuse, or reconstitute existing sites. In Hong Kong a massive new airport, almost half again as large as Heathrow, has been built on a manmade island. In St. Ours, the French are planning to build an $87 million theme park focused on volcanoes.[5] "Vulcania" will include a manmade cone and crater that can simulate eruptions, complete with appropriate sound and light effects and simulated lava flows. Many of Vulcania's exhibits will be underground. The site is not entirely a simulation, however, for it lies in a zone of extinct volcanoes. The surrounding rock is quite genuine basalt, granite, and pumice, and there are real dormant craters nearby. In this realm of fake working volcanoes in a real volcanic zone and real airports on artificial islands, eighteenth-century aesthetic ideas need considerable rethinking. In a characteristic pastiche of references, Vulcania's architect, Hans Hollein of Vienna, claims to be inspired by medieval poetry and early science fiction: Dante's *Inferno* and Jules Verne's *Journey to the Center of the Earth.* Hollein will relandscape an actual extinct volcanic landscape to create a simulated active volcano, using an aesthetic that recycles ideas from science, fantasy, and religion.

With such projects in the offing, it is little wonder that every essay in this volume deals with a subject that does not fit neatly within the traditional categories of the sublime, the beautiful, and the picturesque. All are written with a knowledge of these terms, as deployed by Edmund Burke, Immanuel Kant, and other thinkers since the eighteenth century. But each essayist in his or her own way confronts the complex experience of modern landscapes, which are ineluctably technological, from the high-tech farm to genetically engineered plants to the simulated volcano. Their new terms of analysis include *ephemeral, recycling, off-site, intertext,* and *heterotopia.* As this terminology suggests, rather than seeing landscape primarily as space, these authors understand landscape as a process embedded in narrative, or time. And because these authors see landscapes in time, narration becomes a central concern. They are acutely aware that whether one is concerned with Victorian Manchester, the South of the New Deal, the Nazi Autobahn, a naturalist's description of the New World, or contemporary Louisiana, the question of who is making the narrative of a place is just as important as who constructs the physical landscape. Is the Grand Canyon described as a profitless locality that cannot support agriculture, a sublime wonder, or a potential mining site? Is the smoke of an industrial city construed as pollution or a heartening sign of prosperity? Is an exhausted open pit mine filled with water to be understood as a ravaged landscape, an industrial tourist site, or a recreation area? Is the Appalachian Trail a means to revivify the rural economy or a way to escape into a rugged wilderness experience? These are not necessarily differences between people, for frequently the divisions appear within an individual conscience. For people commonly perceive a single site in contradictory ways. Like one of the early explorers of the Grand Canyon, they may find a landscape sublime and yet wish to build a railroad through it.

Organizations also disagree about schemes to preserve or enhance or recycle particular landscapes. Both Britain and the United States during the 1930s set up organizations to modernize rural areas, yet many of those engaged in the work wished to preserve the landscape and the reassuring values it seemed to embody. As Christopher Bailey puts it, in Britain, "the promise of progress was steadily drained from rural communities in the interwar period, leaving a space in which the preservationists could inscribe their own meanings." In the United States, as Stuart Kidd reports, photographers from the Farm Security Administration "avoided the picturesque" but nevertheless, "they produced many images that dignified the very conditions the FSA was attempting to eradicate." For these image makers, "the landscape of the South was a stimulating antidote to the predictable, anonymous, and modernized cities" of the North. In the same years, Thomas Zeller shows, German Nazis built the autobahn system, with the goal of integrating it harmoniously into the countryside. Amid talk of cleansing rural areas of "alien" plants, the chief of the highway program resisted pressure from landscape architects to follow the natural contours of the land and build in sweeping curves. But he felt the need to justify straight roads by using his own metaphors drawn from nature: "The car is not a rabbit or a deer that jumps around in sweeping lines. . . . Rather, the car resembles a dragon fly or any other jumping animal that moves shorter distances in straight lines and then changes its direction at different points." The tension between modern technology and an idealized nature was a constant, even if it played itself out quite differently in Britain, the United States, and Germany.

How such tensions are embodied in narrative is the central subject of Stephen Mosley's discussion of air pollution in nineteenth-century Manchester. There, a majority of the working class and the industrialists not only accepted but celebrated smokestacks belching smoke in narratives of prosperity. A minority struggled to define a counternarrative that would convince the public otherwise: they not only found it difficult but had to adopt economic and utilitarian arguments in order to get a hearing. Barbara Allen focuses on how managers from chemical factories and local residents along the Mississippi River near New Orleans constitute two interpretative communities. One presents its case to federal regulators in scientific language using computer models about the meanings of the toxic substances that permeate local air, ground, and water. The local African American community tells in vernacular language its stories of odd smells, peculiar feelings, and psychic distress, in the area they call "Cancer Alley."

Most of the examples mentioned thus far deal with the period after industrialization. A brief reminder of the long view is necessary to situate these concerns in larger history, even if one can only suggest the broad outlines. The notion that the world is made up of landscapes has not always been with us, but emerged at a distinct historical moment, roughly two generations before Columbus's discovery of the western hemisphere. Indeed, the term "landscape" can hardly be considered neutral or merely descriptive; in the most literal sense it emerged as part of a distinctive worldview. Partially expressed in the new genre of landscape painting, this worldview coincided with European imperialism, the growth of cities, improvements in agriculture, and the revival of classical learning. In other words, the idea of landscape emerged at a moment of secularization and expansion. A landscape almost by definition was not sacred, uncultivated, or wild, but secular, cultivated, and improved. To value and appreciate a landscape also meant to valorize the labor of figures in the fields and woods. From such scenes developed the aesthetic of the beautiful, as later defined by Burke, Kant, and many other authors.

The appreciation of uncultivated places, such as mountains, developed only several centuries later. The intellectual underpinnings that would later support organizations such as the Sierra Club began to emerge in the seventeenth century, when Europeans started to see as sublime the places akin to the "wilderness" areas of today. Before then, mountains were thought to be ugly deformities of a fallen world, whose surface had been smooth at the creation. The fertile plains were Edenic, while, as Marjorie Nicholson explains, "Mountains were warts, blisters, impostumes, when they were not the rubbish of the earth, swept away by the careful housewife Nature—waste places of the world, with little meaning and less charm." This is not the place to recapitulate her book explaining how mountains became glorious to the imagination, but it may be noted that by the eighteenth century, most writers on aesthetics agreed that the sublime was a fundamental category and found that the most important "stimulus to the Sublime lay in vast objects of Nature—mountains and oceans, stars and cosmic space—all reflecting the glory of Deity."[6]

If the experience of the cosmic in nature was primary, the sublime might also be found in contemplating some human creations, including at first large buildings and structures such as bridges and eventually railroads, factories, and dams. On both sides of the Atlantic the technological sublime emerged alongside the natural sublime, and became particularly popular in the United States.[7] Great works of architecture and engineering, like mountains, vast waterfalls, and canyons, could leave a visitor dumbfounded, amazed, and deeply impressed either by natural forces or by humans' ingenuity in overcoming them. Nor was this impulse limited to modern times, as some ancient authors praised Roman roads, aqueducts, and other engineering projects in similar terms.[8] In short, the idea of the sublime, while conventionally understood in opposition to the beautiful, in fact shares an important affinity with it. Both terms have long been used to describe the ways that human beings shape their landscapes.

Technology is not alien to landscape, but integral to it. Some peoples understand this quite readily, notably the Dutch. However, Americans have long tried not to see this. As is well known, when white Americans encountered an area where Native Americans had hunted and farmed for millennia, it often seemed to them to be untouched. The New World often appeared to them a fragment of Eden that had survived. This perception lead to writing about the land in terms of paradise and the fall,[9] and since at least the eighteenth century some Americans have seen the New World in terms of a prelapsarian garden that is invaded by machines. To take just one example, Daniel Drake, a Cincinnati physician, wrote in 1834 that

> Civilization is a transforming power, and wherever its wand is raised, the surface of the earth assumes a new aspect. The native trees, cut down and consumed, are replaced by the apple and orange; the wild grape, which united their limbs, is succeeded by an exotic, resting on trestles; the rivers are constrained within narrower channels, or turned into canals; and the mossy rocks of their margins are broken with the sledge or exploded with gunpowder; hills are levelled and valleys filled up; a mecadamized road usurps the bed of the little brook, and the rumbling of the coach wheel falls upon the ear, instead of the soft music of its rippling waters; fields of wheat undulate, where the prairie grass waved before, and tobacco and cotton are nourished on the wreck of the cane-brake, which formerly spread its green leaves over the snows of winter. Thus the teeming and beautiful landscape of nature, fades away like a dream of poetry, and is followed by the useful but awkward creations of art.[10]

Drake had witnessed the rapid settlement of the Ohio Valley and could contrast the region's appearance in 1800 with the extensive farms and cities established during his three decades of residence. Already in 1830 he laments the lost "teeming and beautiful landscape of nature" while painting an image of apparent virginal perfection that has given way to the "awkward creations" of civilization. Characteristically, he does not use the word technology. Instead, he speaks of the new cultivated landscapes as "creations of art."

Such passages are about both a present and an idealized memory of a world before that present. The transformations of the land were certainly real enough, as people used axes, sledges, gunpowder, and farm equipment to create a new rural life. Yet where did the idea for this new agricultural world come from, but out of the memories of the immigrants into the region? And when the rural world of horses and small farms they created was itself displaced, the Ohio of the 1830s would in turn become an idyll, a lost "teeming and beautiful" landscape. Reading Drake today, one experiences a double displacement from both his present and his past. Either his old (1800) or his new (1830) landscape might be idealized, and in historical restorations they are.

Displacements may be spatial as well as temporal. Tourists visit and value highly landscapes that embody contradictory aesthetics. In a single day they can pass from Las Vegas to the Grand Canyon, or from the artificial Lake Powell to the natural Rainbow Bridge, and appreciate both sites. Drake was doing something similar in his imagination, for he certainly appreciated the coming of civilization to the Ohio Valley, even though he regretted that an earlier landscape was passing away. Today, the appreciation of (and the consequent attempts to preserve) diverse forms of landscape registers a multiplicity and complexity in memory. Wilderness areas are meant to embody and recall an imagined world before humankind entered the scene. In the United States, Colonial Williamsburg in Virginia, Sturbridge Village and Lowell's restored factories in Massachusetts, and a host of other sites seek to recapture other distinct historical moments through reconstruction. As Schama puts it, "Once a certain idea of landscape, a myth, a vision, establishes itself in an actual place, it has a peculiar way of muddling categories, of making metaphors more real than their referents; of becoming, in fact, part of the scenery."[11] For example, Mark Luccarelli's report in this volume on the invention of the Appalachian Trail shows how the original vision of rejuvenating remote rural economies, a kind of progressive pastoralism, was overwhelmed by a vision of wilderness experience.

In the postmodern world, attempts to construct and preserve landscapes and heritage sites are multiplying. In the United States, there were twenty-five hundred historic preservation groups in 1966 and more than six-thousand a decade later, and the number has not declined since. Historic buildings are renovated with the ostensible goal of saving fragments of the past, but often the function of the building and the clientele change completely in the process. For example, Lowell revitalized its downtown by turning its empty textile mills into an industrial park and historic district administered by the National Park Service.[12] Conversely, as Pia Maria Ahlbäck analyzes, the English city of Wigan has rewritten George Orwell's depiction of it as an oppressive industrial landscape by creating a sleek modern factory surrounded by restored buildings, including a pub named for Orwell. New categories continually arise, based on images from earlier time periods. One might say that there is a "popular culture landscape" of the 1950s, which includes cars with large tail fins, new suburban tracts, split-rail fences, power lawn mowers, and so on, in contrast to a quite different imagined landscape of the 1960s counterculture. This continual reimagination of space entails a continual expansion of heritage that can be better understood by considering more closely the meanings of the two terms *technology* and *landscape*.

One cannot trace an uninterrupted discourse about the relationship of technology and landscape from antiquity to the present. Indeed, the very terms did not exist in anything like their modern meaning until at least the nineteenth century. Both *landscape* and *technology* each have a rich history, and both words have been used more widely, and with more intended meanings in the last generation or so than ever before. *Landscape* is by far the older term, and can be traced in Germanic languages to the twelfth century. As early as 1121 the word *Landschaft* was used to refer to inhabitants of a legally defined zone or region. As Kenneth Olwig has pointed out, the term "literally refers to the shape the land is in with respect to its customs, the material forms generated by those customs, and the shape of the bodies which generate and formalize those customs as law."[13] *Landschaft* only later became linked with the new genre of landscape painting that appeared around 1420, and it was fully integrated into the vocabulary of aesthetics by the eighteenth century. As the term developed, however, the original connection between a people, their culture, and their imprint on physical space gradually attenuated. Landscape increasingly was given a largely aesthetic meaning that was detached from the life and work that a particular space embodied and implied. In the fine arts, the word ceased to imply immediately the agricultural techniques, or the technologies, of a local people.

The visual conventions we now use to understand space were codified in painting, widely disseminated by photography, and further popularized by advertising and film. As Estelle Jussim and Elizabeth Lindquist-Cock argue in *Landscape as Photograph,* "Landscape encompasses both scenery and environment but is equivalent to neither . . . landscape construed as the phenomenological world does not exist; landscape can only be symbolic."[14] Even if a landscape could be found that was an entirely unspoiled natural view, it would be seen through the lens of a powerful visual culture. Not only is there no such thing as natural landscape, but there is no innocent eye to look at it.

The visualization of landscape carries with it a set of interlocking notions, including the long view constructed in rectilinear perspective and framed as a unit. These lines of perspective express an abstract geometrical idea of space that requires an observer to stand outside the scene and frame it in imagination. We stand back, outside, in order to appreciate. The imposition of perspective and the social construction of an external vantage point have some affinities with the attitude and point of view of an engineer. This affinity is by no means surprising, being just one more reminder that until the nineteenth century, the arts were commonly divided into the "fine" and the "useful." The painter depicted what the landscape gardener, the architect, and the engineer constructed.

The term *technology* emerged during the seventeenth century from modern Latin into English, to describe a systematic study of the arts. By the early eighteenth century, a characteristic definition was "a description of the arts, especially the mechanical."[15] The word was not yet widely used. In the United States, the term began to become more familiar after the publication of *Elements of Technology* in 1832 by a Harvard University professor, Jacob Bigelow, and it gradually became a common term.[16] As Leo Marx recently observed, "At the time of the Industrial Revolution, and through most of the nineteenth century, the word *technology* primarily referred to a kind of book; except for a few lexical pioneers, it was not until the turn of this century that sophisticated writers like Thorstein Veblen began to use the word to mean the mechanic arts collectively. But that sense of the word did not gain wide currency until after World War I."[17] By the end of the nineteenth century, the term was embedded in the names of prominent educational institutions, such as the Massachusetts Institute of Technology, but it had not yet become a common term in the discussion of industrialization. Certainly by 1900 it was well understood that machines and inventions could be used to produce new spatial formations, such as the electrified city of skyscrapers, the streetcar suburb, or irrigated farmlands in the desert. Yet the term *landscape* was not always applied to these new formations, and the notion that there might be something called a "technological landscape" had not appeared. *Landscape* tended to be associated with the natural, while the machine was an invading presence. In contrast to its fifteenth-century secular meanings, in the twentieth century *landscape* evoked the idea of timeless harmonies. Whereas it had originally meant something akin to the visual form the land assumes when being cultivated, the term had lost its associations with work or with direct involvement in the creation of social space. The earliest meanings of the word *landscape* and the twentieth-century idea of *technology* had some affinities, but the aestheticized idea of landscape, as being any attractive view, could easily seem to be the opposite of all that industrialization stood for.

As already noted, this treatment of technology and landscape as opposites is misleading. Every landscape implies the technologies that produced it. Drake took this for granted in the 1830s. The same process he witnessed had taken place earlier in New England, where farmers had stripped much of Massachusetts and Connecticut of trees by the end of the eighteenth century. The agricultural machinery then available limited the size of family farms, and the weak transportation and distribution network available before canals or railroads restricted crop specialization. The result was a varied pattern of hillside fields dotted with houses that pleased the eyes of many persons, such as Timothy Dwight, the president of Yale University, who wrote appreciatively of the New England landscape.[18] By around 1820 this world seemed stable and even natural. Yet it was a cultural construction soon to be challenged by more productive midwestern agriculture, which produced wheat and corn so cheaply that it was no longer profitable for New England to raise these crops. As farms were slowly abandoned, a heavily wooded landscape reemerged, just the opposite of what was happening in Drake's Midwest. Nor were such interregional effects of technological development limited by national borders. The cultivation of vast wheat farms west of the Mississippi so lowered European grain prices that some regions there confronted changes every bit as drastic as New England. In Denmark, for example, farmers met the challenge of inexpensive American wheat by shifting to dairy products and bacon, transforming the rural landscape in the process. As these examples suggest, technology and landscape are two aspects of the production of social space not only at particular sites, but as part of a larger system of interlinked transformations.

It is in all likelihood impossible to imagine a landscape that is not socially constructed in two senses. First, people have left their physical mark on the land, both intentionally and inadvertently. The Netherlands offers one of the most striking examples, with its system of canals, dikes, and pumping stations that collectively protect whole regions that lie below sea level. Shipping canals cross over roads; paths lead up to the sea; and new forests are planted on polders, or reclaimed land below sea level. To achieve this control, Dutch engineers have reshaped the courses of rivers and streams to such an extent that often it is difficult to say what is natural and what is artificial. On the other side of the Atlantic, for millennia Native Americans hunted, burned, and farmed North America. Europeans did not settle virgin land, although they did transform it considerably more than did the first inhabitants, logging off large areas; introducing new animals and crops; and constructing farms, mines, roads, towns, dams, and canals.

Second, landscape is always socially constructed because we see it through the invention of perspective vision, the technology of photography, the abstractions of maps, and the traditions of landscape painting that have become a central part of Western visual culture. Thus social construction occurs on two levels, material and psychological, with a constant interplay between site and sight. This psychological dimension of landscape makes it a hazardous concept to employ in argument. At one moment one is speaking about spatial arrangements, and at another, about the entire social and mental system within which they are embedded.

As a group, these essays treat technologies as inseparable from the construction and definition of shared social space. They treat landscapes not as static locations, but as changing sites where new meanings are constantly emerging. From this perspective, the terms *landscape* and *technology* imply one another and are best understood in tandem. It will no longer do for art historians, ecologists, and literary critics to treat landscapes as timeless repositories of cultural values. And it will no longer do for historians or geographers of landscape to overlook the processes of invention, perception, and narration of these shared social spaces. As these essays make clear, because landscape and technology are inseparably intertwined, interdisciplinary work is not merely desirable but indispensable, if we are to recover the sense of space as a living tapestry that records human action, experience, and memory.

NOTES

1. John Brinckerhoff Jackson, *Discovering the Vernacular Landscape* (New Haven, Conn.: Yale University Press, 1984), 8.

2. Simon Schama, *Landscape and Memory* (New York: Harper/Collins, 1995), 61.

3. William Cronon, ed., *Uncommon Ground: Rethinking the Human Place in Nature* (New York: Norton, 1995), 25. Many of the contributors to the pres-

ent volume came to Odense and its Center for Man and Nature in January 1997, to present these essays at a joint conference.

4. See Virginia Scott Jenkins, *The Lawn: A History of an American Obsession* (Washington, D.C.: Smithsonian Institution Press, 1994), 9–35 and passim.

5. All details from Craig R. Whitney, "In France, an Erupting Theme Park," *International Tribune*, 30 July 1998, 2. The key promoter is former president Giscard d'Estaing, who helped secure $12.5 million from the European Union for the project and a third again as much from the French government.

6. Marjorie Hope Nicholson, *From Mountain Gloom to Mountain Glory: The Development of the Aesthetics of the Infinite* (Ithaca, N.Y.: Cornell University Press, 1959), 62, 321–323.

7. See David E. Nye, *American Technological Sublime* (Cambridge, Mass.: MIT Press, 1994).

8. Zoja Pavlovskis, *Man in an Artificial Landscape: The Marvels of Civilization in Imperial Roman Literature* (Leiden, The Netherlands: Mnemosyne, Biblioteca Classica Batava, 1973).

9. Leo Marx traced this theme in the opening chapters of his classic *The Machine in the Garden* (New York: Oxford University Press, 1965).

10. Daniel Drake, *Discourse on the History, Character and Prospects of the West*, with an introduction by Perry Miller (Gainesville, Fla.: Scholars Facsimiles & Reprints, 1955), 17.

11. Schama, *Landscape and Memory*, 61.

12. On historical preservation, see Michael Wallace, "Reflections on the History of Historic Preservation," in *Presenting the Past: Essays on History and the Public*, ed. Susan Porter Bensen, Stephen Brier, and Roy Rosenzweig (Philadelphia: Temple University Press, 1986), 189.

13. Kenneth R. Olwig, "Recovering the Substantive Nature of Landscape," unpublished manuscript, 1996.

14. See Estelle Jussim and Elizabeth Lindquist-Cock, *Landscape as Photograph* (New Haven, Conn.: Yale University Press, 1985), xiv.

15. Raymond Williams, *Keywords*, 2d ed. (London: Fontana, 1983), 315.

16. Jacob Bigelow, *Elements of Technology* (Boston: 1828).

17. Leo Marx, "Technology: The Emergence of a Hazardous Concept," *Social Research*, 64 (1997): 966–967.

18. For discussion of Dwight, see Jackson, *Discovering the Vernacular Landscape*, 59–64.

EPHEMERAL LANDSCAPES

1

Agricultural Technology and the Ephemeral Landscape

PAUL BRASSLEY

In the late 1970s, I took a photograph of a field containing several lines of small stacks of straw bales (Figure 1.1). Without the bales, the field and its surrounding landscape would have been unremarkable; with them, it made an interesting photograph. Nearly twenty years later, some of the reasons for this had become a little clearer to me, and they form the subject of this paper. The bales are just one example of an enormous variety of natural and human-induced changes that affect the appearance of the landscape from minute to minute, or month to month. Within the permanent structure of the landscape, the *ephemeral* landscape is more or less constantly changing. It is a vital but often unrecognized component of the landscape as a whole.

It is indisputable that technical change produces landscape change,

---

I took the photograph in this essay as a result of a remark by Dr. Bill Slee about the importance of ephemeral objects in the landscape. Having taken it, I thought no more about it until the spring of 1996, when I began to develop the arguments contained in this paper. For numerous subsequent discussions I am grateful to Angela Brassley, Geoff Hearnden, Alan Bruford, Christopher Bailey, Irene Klaver, and my colleagues at Seale-Hayne: Sue Blackburn, Martyn Warren, Derek Shepherd, Ian Whitehead, Ian Kemp, Peter Holgate, Graham Busby, and, especially, Anne Millman. I am also grateful to Svend-Erik Larsen and David Nye of Odense University. Had they not arranged the conference "Landscape and Technology" in January 1997, this paper would never have been written.

The essay represents work in progress, rather than a completely worked out and finally articulated argument. I am still at the stage of wondering whether the concept of ephemera is so obvious as to be hardly worth writing about or of great but hitherto unrecognized significance. However, responses so far have been sufficiently positive to encourage further work.

FIGURE 1.1 Small stacks of straw bales in the late 1970s. Courtesy of the author.

and it follows that changes in agricultural technologies produce changes in agricultural landscapes. It can be demonstrated that the speed of change has increased since 1950. These assertions are briefly justified in this essay, but the main point of the argument is to suggest that the importance of the ephemeral landscape, and the impact of technical change on it, has been largely unrecognized in the landscape literature and among landscape professionals. This is unfortunate, because ephemeral changes affect landscape perceptions, are readily appreciated by ordinary people, and form an important component of the response to landscape by artists in many different media.

Since the development of agriculture in the Neolithic period, changes in agricultural technology have transformed the rural landscapes of the world. In this paper, I focus on the highly cultivated western European landscape, but comparable changes have obviously occurred in the North American and Australian landscapes.[1] For present purposes, these changes might be crudely grouped into four periods (in Europe—the periodization would be different on other continents):

Changes during the shift from hunter-gatherer to neolithic agricultural societies. In this period, the introduction of agriculture resulted in the replacement of woodland-dominated landscapes by cultivation-dominated landscapes, with the concurrent development of new myths to explain the relationship between people and nature.

Prehistoric to medieval changes, such as the introduction of the heavier plough, which allowed the expansion of cultivation, new crops, and the replacement of block fields by common fields.

Changes between the medieval period and the mid-twentieth century, such as the development of the horse-hoe, requiring crops to be grown in rows, and the drill, which produced the rows; the introduction of new crops, such as the potato; and the process of enclosure.

Changes since 1950, including hedgerow removal; the replacement of the binder by the combine harvester, of haycocks by bales and then by silage, and of the horse-hoe by the sprayer; and the reintroduction of old crops such as oilseed rape and linseed.[2]

The speed with which these changes have occurred has increased in the twentieth century, and especially since 1950. Oliver Rackham has argued that photographs of southeastern England, taken by the Luftwaffe in 1940 for purposes other than historical research, record "what was still, in many places, a medieval landscape, much of it since damaged or effaced."[3] The impacts of the chainsaw on woodland, and of earth-moving machines on hedgerows are well known. But the same sort of thing was going on *within* the fields surrounded by the hedges: the rate of adoption of new technology has increased since World War II. In contrast to the gradual introduction of agriculture itself, over several thousand years, or the spread of the seed drill, which took several hundred years, the replacement of the binder by the combine harvester and the associated baler in the United Kingdom was much more rapid (Table 1.1). In the twenty years between 1950 and 1970 the combine replaced the binder almost completely, and with the binder went the sheaves and stooks that had been part of the harvest landscape for hundreds, if not thousands, of years. Similarly, between 1970 and 1990 in

TABLE 1.1 Changes in Cereal Harvesting Machinery in the United Kingdom, 1942 to 1980, in Thousands of Machines[4]

	Binders	Combines	Balers
1942	102	1	No data
1946	119	3	No data
1950	120	10	16
1960	75	48	58
1971	0	57	70
1980	0	47	74

the United Kingdom, silage replaced hay as the main method of fodder conservation, and the hayfields and their associated flora and fauna became much rarer.[5] These changes and many more, discussed below, have had dramatic effects on the landscape.

The impact of change on human perceptions of landscape is revealed more clearly if we introduce a new conceptual dichotomy: *ephemeral* landscape as opposed to *permanent* landscape. Permanent components of the landscape are those that do not change in the short term and may change only slowly in the long term. Buildings and other aspects of settlements, field boundaries, woods and forests, roads, railways, and canals are all landscape components that fit into this permanent category. Clearly, they are not immune to change. Old brick is replaced by concrete; hedges are replaced by barbed wire or removed to make fields bigger; broad-leaved woodlands are cut down or replaced by conifers; and settlements expand or, more rarely, contract. Such changes to the permanent landscape are much discussed, readily identified by landscape professionals, and often the subject, in the United Kingdom at least, of planning or land use legislation or listed building consent. Until recently hedges have been an exception to this rule, but there are currently (in 1997) proposals to introduce legislation to control hedgerow removal.[6] I do not address these changes here, except to note that they affect the landscape and are much discussed in its literature.

The ephemeral landscape, in contrast, is that which changes more or less rapidly. In a sense, all human life is ephemeral, as A. S. Byatt points out: "The individual appears for an instant, joins the community of thought, modifies it and dies; but the species, that dies not, reaps the fruit of his ephemeral existence."[7] Ephemeral landscape effects range from the momentary influences of cloud and sun, gone before the camera can be removed from its case, to the development of the seasons. The vast range of ephemeral effects may be divided into three broad categories: natural, human induced, and a mixture of the two.

Natural ephemera are the results of the interactive changes of weather and the seasons. As the clouds part, and the sun shines on water, or through trees, or on distant hills, and cloud shadows follow each other across the fields, the landscape may be transformed within seconds. Much of what the eye sees of the great wide landscapes of lowland plains such as the English Fenland is the sky. Twice a day the landscapes of the estuary and the seashore change as the tide comes in, covering rocky ledges, or goes out, exposing mudflats and sandbanks to wading birds and replacing a scene of rippling wavelets by one of shining silt. Over a longer time scale, a light snowfall changes the greens and browns of the rural winter landscape to dark lines and patches on a white background. Summer drought and ripening corn turn the greens of spring to ochre or sienna. The grass of old hay meadows was dotted with the whites, yellows, and pinks of spring flowers, which is perhaps one reason why the few remaining unsprayed and unfertilized meadows are so carefully conserved. The colors of the trees in the fall in New England attract thousands of visitors each year, as do the spring bulb fields in The Netherlands. The plants that cover the soil and so produce the colors and textures of the natural landscape have their times of growth and decay, blossom and fruit. For most, the cycle lasts a few months, but for some it lasts only a few days and for others many years. Taken together, the state of the natural landscape is generally one of constant flux.

Animals, too, experience ephemeral change. Indeed, the mayflies take their zoological name—*Ephemeroptera*—from the same Greek root as the term itself: *ephemeros,* which means 'living for a day'.[8] Mountain hares and stoats change their coats in the winter; the sexual activity of sheep is affected by changing day length; and, as Chaucer pointed out in the late fourteenth century, "smale fowles maken melodye" in the spring,

> Whan that Aprille with his shoures sote
> The droghte of March hath perced to the rote.

And without the song of the birds and the buzz of the insects, the awakening experience of the spring would be diminished, as Rachel Carson so influentially argued.[9]

Human-induced ephemera are usually associated with agriculture, principally because agriculture is the major land use in Europe.[10] The choice of crop and the method of its cultivation and harvesting can all affect the appearance of the landscape. Since the 1970s the impact on the landscape of the particularly violent yellow of oilseed rape crops in flower has attracted much adverse comment; equally, many people have found the soft blue of the more recently popularized linseed an attractive landscape feature:

> A seven-acre field had suddenly become tinctured with the colour of Mediterranean skies. It happened almost in a night. One day there was a faint azure mist upon the field, like smoke from a squitch-fire. Next morning when the sun came up a cerulean carpet covered it; and we almost caught our breath at the sight of this miracle, . . . Now, between the orchards and the corn, appeared this astonishing lagoon of blue which caught and held the eye so that within an hour the whole neighbourhood was talking about it. "Have you seen old William's field of linseed?" people said. "It does your heart good to look at it."[11]

Both oilseed rape and linseed owe their present extent to the level of subsidy they receive under the terms of the European Union's Common Agricultural Policy, which also produces something that is not a crop but affects the appearance of the landscape: set-aside land.[12]

The cultivations carried out in the process of farming also change the way the land looks. Ploughing a grass field changes a uniform green to a corduroy brown. It will stay that way for several months over the winter if it is not sown with a corn crop until spring. Then it will gradually green over again as the young corn sprouts and shoots. In recent years the popularity of winter-sown cereals has increased, so the brown phase is shorter and the green longer, and the amount of ephemeral variation is thus reduced. Other technical changes have also had their ephemeral effects. Henry Rider Haggard, the writer and farmer who traveled from Baldock to Ashwell in Hertfordshire at the beginning of the twentieth century, noted a field of wheat yellow with charlock and a cornfield "stained blood-red with poppies."[13] Hand weeding and horse-hoeing were not always very effective. Today, of course, both of these weeds are controlled by the use of selective herbicides, and so the corn crops are a uniform green, with no splashes of yellow or red. Even the shade of green has been changed—less yellow in it, and more dark bluey-green—by the increased use of nitrogenous fertilizers. Farmyard manure is still spread on the land, although sixty years ago heaps of manure would be piled in rows from a horse-drawn dung cart, to be spread about afterwards by a man with a fork, whereas now slurry is spread directly from a muckspreader pulled by a tractor. Indeed, it might be argued that the replacement of the horse by the tractor, and of muscle power in general by machinery, affects the whole rural environment, not only visually, but also in terms of the characteristic noises and smells now associated with rural life. But still rolling grassland in spring produces, for just a few days, the lawnlike stripes of alternating dark and lighter greens.

Methods of harvesting, whether of grass or corn, have also been radically altered, with concomitant effects on the ephemeral landscape. For centuries grass was conserved as hay, cut with the scythe, spread and tossed with pitchforks, raked up, made into haycocks, and finally carted, all by the muscle power of as many men and women as could be found.[14] Then a series of new machines, first the horse-drawn mowers and hay tedders that appeared in the nineteenth century, and later the pick-up baler converted the hay harvest landscape from one of loose hay and many people to one of windrows, followed by piles of bales, and many fewer people. In the last twenty years, the scene has changed again, as the bulk of the grass crop in the United Kingdom is now conserved as silage, so the windrows of drying grass need not be left so long to dry, and the piles of small bales have disappeared completely or been replaced by big bales wrapped in black polythene.[15] Similarly, the appearance of the countryside during the corn harvest has been fundamentally altered by the demise of the binder and the rise of the combine harvester. Instead of the lines of stooks that covered the cornfields, often for several weeks in August and September, the landscape change that signals the progress of the harvest is the replacement of the standing corn by the swaths of straw waiting to be baled, followed by the bales waiting to be carted.[16]

Livestock farming has also been subject to technical change, and it, too, has had its effect on the ephemeral landscape. Milking bails (small portable buildings in which cows were milked in the fields) have now virtually disappeared, having once been a common sight in the fields in some districts. Free-range hens and outdoor pigs went, too, although in recent years some have reappeared. Lambs, which once gamboled in the spring, are now a common sight in the southwest of England from Christmas onwards. The change in the principal dairy breed over the last sixty years in the United Kingdom means that brown Shorthorns have largely been replaced by black-and-white Friesians. Does this, too, not affect the appearance of the landscape in which they graze?

Simply because agriculture is the dominant land use, agricultural change has had the greatest impact on the ephemeral landscape of Europe in the present century. It is not unique in this effect, however, Woodland management methods have also changed fundamentally in the last century or two and consequently have affected the ephemeral landscape. Coppicing was the most common form of medieval woodland management, and although in the United Kingdom it was clearly in decline in the second half of the nineteenth century, significant areas of woodland continued to be worked on coppice rotations until the early years of the twentieth century. The technique of coppicing, in which the underwood was regularly cut and allowed to regrow over a regular cycle (varying between seven and twenty or more years) meant that the appearance of the woodland changed over the cycle: it was open to the light just after cutting, so that the woodland floor in spring would be covered by bluebells or primroses, and darkened as the new crop grew and the canopy closed over (all in addition, of course, to the annual cycle of new leaf in the spring and leaf fall in the autumn). Aurally, the thwack of the woodsman's axe has been replaced by the scream of the chainsaw. Streamside willows were regularly pollarded for basket-making, so that the appearance of the tree changed from year to year. In other times and other continents ephemeral change in the woodland must have been familiar. Native Americans and Australian Aborigines both managed woodland by burning: the idea of the unchanging and impenetrable primeval forest has been demonstrated to be a misconception.[17]

Thus we can find numerous landscape features that are ephemeral, some natural, some produced by human activity, and a few that are mixtures of the two (such as the bleached appearance of grass fields when drought follows the silage cut). And the human-induced ephemeral landscapes have been and are being altered by technical change, especially but not exclusively in agriculture. Does this matter, except in an antiquarian sense?

Landscape ephemera are important because human perceptions and evaluations of landscapes are sensitive to change. At the very least, it might be argued that ephemeral changes are more noticeable because the eye slides over that which is a permanent part of the landscape but is arrested by that which has changed. In some ways, urban people may be more sensitive to rural landscape changes because they see the rural landscape intermittently. This line of reasoning might be extended to include the changes we *expect* to happen: we expect the appearance of leaves in the spring and wintery weather at Christmas, and if spring is late or early, or Christmas "unseasonably warm," we remark upon it.[18] Similarly, a dry brown English summer landscape provokes comment. This expected change might be termed the *normative ephemeral.*[19] It is perhaps a special case of the more general concept of ephemeral change as an extra dimension or component in the landscape: if rural landscapes are constructed of several components—fields and hedges, woods, buildings and settlements, and so on—then the ephemeral changes that affect the appearance of the fields, or the color of the hedges and the woodland, may be seen as another component.

So far I have relied on anecdote or appeals to personal experience. This is not necessarily a weakness, because quite clearly each year several thousand people vacation in New England in the fall, when the colors of the vegetation are at their most magnificent, or watch the sun set over the Grand Canyon, or drive round the Dutch bulb fields. In other words, the desire to experience ephemeral landscape effects persuades people to part with disposable income. Perhaps more importantly, this behavior appears to be consistent with generally accepted theories of landscape perception.

Theories or explanations of landscape perception and preference might be divided, insofar as they may be categorized at all, into two schools: the biological or atavistic and the cultural or symbolic. The atavistic explanation of landscape preference is based on the argument that humans are biologically programmed, as a result of the thousands of years that their ancestors spent in the Paleolithic stage of development, to act, albeit unconsciously, as though they were hunter-gatherers. Thus a landscape that affords a good view over a wide area from a hidden vantage point and contains no potential hazards is preferred to a landscape in which it is difficult to detect the approach of a threat and there are no potential hiding places. This is a crude characterization of Appleton's prospect-refuge theory, which is based on the logical argument that there are Darwinian survival advantages in feeling comfortable in the sort of landscape with prospects and refuges, and vice versa. Clearly it is atavistic in arguing that the psychology of present-day people will be affected or determined by the preferences of their Paleolithic ancestors.[20]

If this approach is correct, can it explain or encompass the importance of the ephemeral? If landscape preferences are related to what was important for the survival of Paleolithic people, it would be useful to be able to demonstrate that prehistoric people were used to a landscape that changed, perhaps in predictable ways. In practice, much would have depended on the local environment. Tuan outlines the differences between the perceptions of the BaMbuti pygmies of the Congo rainforest and the Gikwe and !Kung bushmen of the Kalahari Desert. In the rainforest, seasonal variation is minimal, the sun is largely hidden by the trees, and the BaMbuti have little perception of seasonal change. In the desert, visual acuity is vital for tracking game animals and locating edible fruits and roots.[21] In these circumstances it seems likely that experience of the ephemeral aspects of the landscape would have played an important role in assessing its capacity to provide the means of subsistence for Paleolithic people. In the absence of watches, calendars, and supermarkets, close observation of ephemeral changes might well be associated with fitness for survival. The writer Bruce Chatwin, in his travel pieces, repeatedly argues for the naturalness of the itinerant way of life, claiming that American scientists used encephalographic studies to demonstrate that changes of scenery stimulate the brain.[22] Thus an atavistic approach to landscape preference would appear to support the significance of ephemera.[23] Using the same basic approach, Orians points out that human response to pattern variation—and could not many ephemeral changes be so described?—might be termed stimulus discrepancy: "People often respond with pleasurable emotions to minor variations in a familiar stimulus pattern, but with negative feelings toward major deviations from the same pattern."[24] Similarly, Kaplan and Kaplan argue that one major factor accounting for otherwise unexplained preference differences in their experiments is familiarity, and this would presumably explain why some people might feel antipathetic or even antagonistic toward new ephemera: "People who share an environment would presumably have greater familiarity with its vegetational patterns and seasonal effects, and this would be expected to influence their perceptions and preferences."[25]

The alternative school of thought emphasizes the cultural or symbolic importance of landscapes, whereby "a landscape is a cultural image, a pictorial way of representing, structuring or symbolising surroundings," as Daniels and Cosgrove note. This vision of landscape or, perhaps more accurately, of representations of landscape, emphasizes its textual aspects. The landscape itself is associated with its meaning, its symbolic significance, but in the shifting (ephemeral?) perspectives of postmodernity, the meanings resist any final definition. As Daniels and Cosgrove conclude:

> From such a post-modern perspective landscape seems less like a palimpsest whose "real" or "authentic" meanings can somehow be recovered with the correct techniques, theories or ideologies, than a flickering text displayed on the word-processor screen whose meaning can be created, extended, altered, elaborated and finally obliterated by the merest touch of a button.[26]

Nevertheless, the problem of preference remains. Can this approach tell us anything about why people appear to prefer some landscapes to others? The level of opposition to the "sitka slums" produced by modern commercial forestry in the United Kingdom suggests that the people who live there prefer trees that change with the seasons. Would the same apply to Germans brought up in the Black Forest? The emphasis on cultural symbols presumably suggests that there may be some landscapes, or landscape components, that have an iconographic significance. Could we therefore argue that sheaves of corn or horse-ploughed fields have symbolic meanings different from those of trailer-loads of grain and tractor-ploughed fields?[27] Prince argues that landscape art tends to be nostalgic about agriculture, and de Vries and Klaver demonstrate that Dutch painters tended to omit the latest technology of their own day from their landscapes.[28] Ephemeral components of the landscape, it would seem, can have just as much symbolic significance as hedges, trees, or any other component of the permanent landscape. Perhaps we could go so far as to argue that the loss of public sympathy for agriculture in the late twentieth century came about because it began to symbolize the new and not the traditional. And many of the outward and visible signs of the new were ephemera, such as weed-free fields, new machines, new harvesting methods, and new crops.[29]

To recapitulate the argument so far, there are ephemeral components in the landscape, and no matter whether their significance is analyzed from an atavistic or a symbolic viewpoint, they appear likely to affect landscape perceptions and preferences. The ephemeral is a significant component of the landscape. Is it really possible to claim that this has never been recognized before?

The impact of technical change on the *permanent* landscape is widely recognized and accepted by policymakers and incorporated into agricultural, environmental, and land use planning policies. Planning permission is required in the United Kingdom to erect new buildings, to change or destroy old buildings that are officially listed as being of architectural or archeological value; and, in some circumstances, to cut down or plant trees. Legislation is currently under discussion to extend the same kind of provisions to hedges, especially those of historical or conservation importance—generally those that have been in the landscape for a long time. But these legislative proposals appear to be silent on the maintenance of the hedges, and it is that which affects their appearance and vigor.[30]

This is typical of the way in which the ephemeral landscape, and the impact of technical change on it, has been largely ignored by policymakers, landscape professionals, and the landscape literature. Consequently, no official permission is required to plant large areas of oilseed rape, linseed, or sunflowers, all of which have dramatic effects on the landscape. The same applies to cultivation, harvesting, and animal husbandry techniques. Landscape evaluation methods such as the Coventry-Solihull-Warwickshire method require the enumeration of features that may be identified from a map and are therefore certain to be permanent. The only exception among all the various techniques that have been suggested is the old Fines method, which recognizes that the most spectacular landscape effects can be produced by particular weather conditions and so are of necessity ephemeral.[31] The same neglect of the ephemeral is apparent among the mainstream landscape perception and preference literature. Although I have sought to demonstrate that the suggested importance of ephemera is not inconsistent with prospect-refuge and other atavistic theories, this does not mean that prospect-refuge theory attaches any importance to ephemeral effects.

The same appears to be true of much of the landscape psychology literature. Bourassa's extensive survey of landscape preference studies fails to reveal any that mention ephemera. Given the preponderance of studies based on the inspection of landscape photographs, this may not be surprising.[32] Indeed, "surprisingness" was one of the variables affecting environmental aesthetics discussed by Wohlwill as long ago as 1976. He concluded that its investigation would require a "temporal aspect" to be built into the methodology, "which has been notably absent in this area, with its almost exclusive resort to photographic slides."[33] Whether or not surprisingness is exactly the same as the ephemeral is an interesting question, but there is perhaps at least some similarity. More recently, Kroh and Gimblett, in a critique of the use of photographs, found that on-site experience of all the elements in the landscape, such as traffic noise, birdsong, and weather conditions [all of them ephemeral], needed to be represented. Hetherington, Daniel, and Brown, using moving and still pictures, found that motion and sound influenced judgments of scenic beauty and concluded that it was possible to overlook important environmental attributes, "because they were not adequately captured by the representation medium used as a surrogate for the environment under investigation." Despite the shortcomings of static photographs, it is interesting to note that in the Kaplans' survey of preference studies, scenes with ephemeral components rated higher than those without and in other studies the preferences elicited appear to indicate that people select landscapes that they intuitively know contain the most potential for ephemeral change: small fields bounded by hedges, broad-leaved trees rather than sitka plantations, and wildflower meadows. Thus there is evidence for the importance of the ephemeral in the landscape literature, but it is not recognized as such.[34]

Beyond the landscape literature, other writers have discussed the significance of the ephemeral or something like it. Harvey, writing on postmodernism, points out its "total acceptance" of ephemerality. Mukarovsky, exploring aesthetics, examined the concept of intentionality and unintentionality in art, and it might be argued that there are at least parallels between this and the permanent/ephemeral dichotomy. Klaver, introducing her concept of indeterminacy in wilderness, suggests that "Wild is what comes and goes," which is clearly not dissimilar to the idea of the ephemeral.[35] With these exceptions, academic writers appear to have avoided ephemera.

The same cannot be said of artists. Some modern landscape artists, such as Richard Long and Andy Goldsworthy, make a virtue of the ephemeral nature of their works, many of which only have a permanent existence in the form of photographs.[36] We might also argue that interest in ephemeral effects in the landscape is not limited to the twentieth century: the sun rising behind Norham Castle one morning in September 1797 and the watercolors he made of it are said to have transformed J. M. W. Turner's career. Similarly, much of Monet's work aims at capturing ephemeral changes in light and atmosphere. Landscape photographers have pointed out how essential it is to wait for the right light to make a picture, and the ephemerality of sand and leaf sculptures has been deliberately emphasized by one advertiser at least.[37]

The concept of ephemera, or the permanent/ephemeral landscape dichotomy, may have more general implications for landscape perception theory and environmental psychology. Much of the literature in these disciplines is concerned with demonstrating that people like nature. Ulrich demonstrated improved recovery rates in mildly stressed persons who were shown "unspectacular nature scenes," compared with those who were shown urban scenes. Kaplan suggests that experiencing nature helps people to recover from the stress of modern living and that "fascination" is a central component in enabling people to solve problems and maintain stability.[38] Much of the Kaplans' work concentrates on spatial variables, but are not time variables also significant? Is it the ephemeral changes that fascinate? Can we go so far as to claim that the reason why people like nature is that it is subject to ephemeral change, to which we are genetically programmed to respond positively? I suggest that we can, but much research will be required before we can be sure.

NOTES

1. See, for example, M. Williams, *Americans and Their Forests: A Historical Geography* (Cambridge: Cambridge University Press, 1989); G. G. Whitney, *From Coastal Wilderness to Fruited Plain: A History of Environmental Change in Temperate North America from 1500 to the Present* (Cambridge: Cambridge

University Press, 1994); N. Barr and J. Cary, *Greening a Brown Land: The Australian Search for Sustainable Land Use* (London: Macmillan, 1992).

2. Examples of myths that explain the relationship between people and nature are the Mesopotamian Gilgamesh story and the Judeo-Christian Garden of Eden story. Nevertheless, there remain many tree-centered creation myths, as outlined in S. Schama, *Landscape and Memory* (London: Harper/Collins, 1995) 82, 218. On changes from the prehistoric to medieval periods, see Grith Lerche, *Ploughing Implements and Tillage Practices in Denmark from the Viking Period to about 1800 Experimentally Substantiated* (Herning, Denmark: Poul Kristensen, 1994); W. G. Hoskins, *The Making of the English Landscape,* 3d ed. (London: Hodder and Stoughton, 1988), 41; R. Bartlett, *The Making of Europe: Conquest, Colonization and Cultural Change, 950–1350* (London: Penguin, 1993), chap. 5 and 6. On changes between the medieval period and the mid-twentieth century, see M. Overton, *Agricultural Revolution in England: The Transformation of the Agrarian Economy 1500–1850* (Cambridge: Cambridge University Press, 1996); K. Ravnkilde, *From Fettered to Free: The Farmer in Denmark's History* (Hurst, Denmark: Danish Language Services, 1989).

3. O. Rackham, *The History of the Countryside* (London: J. M. Dent and Sons, 1986), 22.

4. H. F. Marks and D. K. Britton, *A Hundred Years of British Food and Farming: A Statistical Survey* (London: Taylor & Francis, 1989), 146; Ministry of Agriculture, Fisheries and Food, *Agricultural Statistics 1954/5: England and Wales Agricultural Censuses and Production* (London: HMSO, 1956), 89; Ministry of Agriculture, Fisheries and Food, *Agricultural Statistics 1960/1: England and Wales Agricultural Censuses and Production* (London: HMSO, 1962), 102; Ministry of Agriculture, Fisheries and Food, *Agricultural Statistics England and Wales 1976–7: Agricultural Censuses and Production* (London: HMSO, 1980).

5. Paul Brassley, "Silage in Britain, 1880–1990: The Delayed Adoption of an Innovation," *Agricultural History Review* 44, pt. 1 (1996): 63–87.

6. In the United Kingdom, the Countryside Commission's technique for landscape recording concentrates on permanent landscape components, although it would be unfair to claim that they were totally insensitive to ephemera: one of their official publications quotes an inspector's report on the designation of the North Pennines Area of Outstanding Natural Beauty which includes the sentence "The feel of the wind, the warmth of the sun, the scent of flowers, the glare of snow, or a sudden break in low scudding cloud, all contribute to the perception of landscape." Josephine Meredith, "Beauty and the Eye of the Beholder," *Countryside Commission News* 29 (November/December 1987): 4–5; Department of the Environment, The Welsh Office, and Ministry of Agriculture, Fisheries and Food, "Protection of Important Hedgerows: Joint Consultation Paper" (Department of the Environment, London 1996).

7. A. S. Byatt, *Possession: a Romance* (London: Vintage, 1990), 4.

8. M. Chinery, *A Field Guide to the Insects of Britain and Northern Europe,* (London: Collins, 1973), 58–62.

9. Geoffrey Chaucer, prologue to *The Canterbury Tales,* ed. F. N. Robinson (Oxford: Oxford University Press, 1966), line 102; Rachel Carson, *Silent Spring* (London: Hamish Hamilton, 1963).

10. Agriculture accounts for 43 percent of the land area of the European Union. The next largest land use is woodland, which accounts for 34 percent of the total area. Statistical Office of the European Communities, *Agriculture: Statistical Yearbook 1996* (Brussels: Statistical Office of the European Communities, 1996), 29.

11. John Moore, *The Blue Field* (London: Collins, 1948), 30. Moore's novel is an account of an individual rebellion against state direction of farming during World War II, but his description of the landscape effect of linseed (or flax—the crop is called linseed when it is grown for its oil-bearing seed and flax when it is grown for its fiber, from which linen is made) applies equally well to present-day cultivation of the crop.

12. Linseed and oilseed rape are grown, and land is set aside (i.e., does not have a commercial crop grown on it), in response to grants made under the Common Agricultural Policy of the European Union. See Paul Brassley, "The Common Agricultural Policy of the EU," in *The Agricultural Notebook,* ed. R. J. Soffe (Oxford: Blackwell Science, 1995), 11.

13. H. Rider Haggard, *Rural England* (London: Longmans, Green, 1906), 1:553.

14. An artist's impression of the landscape effects of traditional haymaking methods may be seen in *The Dixton Harvesters,* an anonymous painting of about 1710. Dixton is in Gloucestershire (UK) and the painting hangs in the Cheltenham Art Gallery and Museum. It is discussed further in D. Souden and D. Starkey, *This Land of England* (London: Muller, Blond and White, 1985), 3.

15. Brassley, "Silage in Britain," 86.

16. John Higgs, *The Land* (London: Studio Vista, 1964), plates 229 and 230.

17. On the use of willows for basketmaking, see H. L. Edlin, *Woodland Crafts in Britain,* (London: Batsford, 1949). On the misconception of the primeval forest, see S. C. Bourassa, *The Aesthetics of Landscape* (London: Belhaven, 1991), 69; N. Barr & J. Cary, *Greening a Brown Land: The Australian Search for Sustainable Land Use* (London: Macmillan, 1992); and G. G. Whitney, *From Coastal Wilderness to Fruited Plain: A History of Environmental Change in Temperate North America from 1500 to the Present* (Cambridge: Cambridge University Press, 1994), 107–20.

18. Perhaps for this reason discussion of Christmas in the film *Babe* (Noonan, 1995) against a background of trees in full leaf was confusing to some northern hemisphere audiences; the film was shot in Australia.

19. I am grateful to Professor Jan de Vries for suggesting this term.

20. Jay Appleton, *The Experience of Landscape* (Chichester, England: John Wiley, 1986).

21. Yi-Fu Tuan, *Topophilia: A Study of Environmental Perception, Attitudes, and Values* (Englewood Cliffs, N.J.: Prentice-Hall, 1974), 77–81.

22. This point is made by Blake Morrison in his review of *Anatomy of Rest-*

lessness: Uncollected Writings by Bruce Chatwin, edited by J. Borm and M. Graves, *Independent on Sunday,* 28 July 1996, 29.

23. The atavistic approach is also used by R. R. Abello and F. G. Bernaldez, "Landscape Preference and Personality," *Landscape and Urban Planning,* 13 (1986): 19–28; R. S. Ulrich, "Human Responses to Vegetation and Landscapes," *Landscape and Urban Planning,* 13 (1986): 29–44; S. Kaplan, "Perception and Landscape: Conceptions and Misconceptions," in *Environmental Aesthetics: Theory, Research and Applications,* ed. J. L. Nasar (Cambridge: Cambridge University Press, 1988), 45–55. I am grateful to Anne Millman for these references.

24. G. H. Orians, "An Ecological and Evolutionary Approach to Landscape Aesthetics," in *Landscape Meanings and Values,* ed. E. C. Penning-Rowsell and D. Lowenthal (London: Allen and Unwin, 1986), 19.

25. R. Kaplan and S. Kaplan, *The Experience of Nature: A Psychological Perspective* (Cambridge: Cambridge University Press, 1989), 86.

26. S. Daniels and D. Cosgrove, "Introduction: Iconography and Landscape," in *The Iconography of Landscape,* ed. D. Cosgrove and S. Daniels (Cambridge: Cambridge University Press, 1988), 1, 8.

27. Thomas Hardy used a cultivation image—"Only a man harrowing clods / In a slow silent walk"—as a metaphor for continuity in *In Time of 'The Breaking of Nations',* reprinted in *The New Oxford Book of English Verse,* ed. H. Gardner (Oxford: Oxford University Press, 1972) 771, line 192.

28. H. Prince, "Art and Agrarian Change, 1710–1815," in Cosgrove and Daniels, *op.cit.* 98–118; Jan de Vries, "Making the Dutch Landscape and Learning to Appreciate It," and Irene Klaver, "Landscape and Nostalgia: Dutch Polders" (papers presented at conference, "Landscape and Technology," Odense University, Odense, Denmark, 9–10 January 1997).

29. Ailsa Tanner, a Landgirl (female farm worker) during World War II, made the same point: "There was a lot of hard work involved but the countryside looked so much more attractive at that time with the pattern of stooks and ricks over the fields. Now, great blocks of straw and polythene covered cylinders of hay do not look the same at all, and what is most sad is that so few people are involved in farm work." Joan Mant, *All Muck, No Medals: Landgirls by Landgirls* (Lewes, England: The Book Guild, 1994), 192.

30. See note 6.

31. A. W. Gilg, *Countryside Planning* (London: Methuen, 1978), 211–215; K. D. Fines, "Landscape Evaluation: A Research Project in East Sussex," *Regional Studies* 2 (1968): 41–55. Both of these methods are now outdated, but it is interesting to note that the UK Countryside Commission's Landscape Character Programme, which places greater emphasis on the cultural significance of landscapes, uses mainly permanent landscape features in the lists of Key Characteristics of each of the regional character areas. Countryside Commission, *The New Map of England: A Celebration of the South Western Landscape* (Cheltenham, England: Countryside Commission, 1994).

32. S. C. Bourassa, *The Aesthetics of Landscape* (London: Belhaven, 1991),

chap. 5; R. B. Hull IV and W. P. Stewart, "Validity of Photo-Based Scenic Beauty Judgements," *Journal of Environmental Psychology* 12 (1992): 101–114; A. T. Purcell et al., "Preference or Preferences for Landscape?" *Journal of Environmental Psychology* 14 (1994): 195–209; E. Strumse, "Perceptual Dimensions in the Visual Preferences for Agrarian Landscapes in Western Norway," *Journal of Environmental Psychology* 14 (1994): 281–292; E. Strumse, "Environmental Attributes and the Prediction of Visual Preferences for Agrarian Landscapes in Western Norway," *Journal of Environmental Psychology* 14 (1994): 293–303; K. Korpela and T. Hartig, "Restorative Qualities of Favourite Places," *Journal of Environmental Psychology* 16 (1996): 221–233.

33. J. F. Wohlwill, "Environmental Aesthetics: The Environment as a Source of Affect," in *Human Behavior and the Environment: Advances in Theory and Research,* ed. I. Altmann and J. F. Wohlwill (New York: Plenum Press, 1976), 1:52–53.

34. D. P. Kroh and R. H. Gimblett, "Comparing Live Experience with Pictures in Articulating Landscape Preference," *Landscape Research* 17 (1992): 58–69; J. Hetherington, T. C. Daniel, and T. C. Brown, "Is Motion More Important Than It Sounds: The Medium of Presentation in Environment Perception Research," *Journal of Environmental Psychology* 13 (1993): 283–291; R. Kaplan and S. Kaplan, *The Experience of Nature: A Psychological Approach* (Cambridge: Cambridge University Press, 1989), 42–45; Strumse, "Perceptual Dimensions" and "Environmental Attributes"; T. R. Herzog, "A Cognitive Analysis of Preference for Field and Forest Environments," *Landscape Research* 9 (1982): 10–16; A. J. Brown, "Impact of Enhanced Biological Resources on Landscape Value for Tourism: Exploring the Perception of Biodiversity through the Medium of Small Groups," Institute of Grassland and Environmental Research, North Wyke, Devon, England, 1995. I am most grateful to Anne Millman for these references.

35. David Harvey, *The Condition of Postmodernity: An Enquiry into the Origins of Cultural Change* (Oxford: Blackwell, 1990), 94. I am grateful to Sue Blackburn for this reference. Jan Mukarovsky, *Structure, Sign, and Function* (New Haven, Conn.: Yale University Press, 1978), chap. 8. I am grateful to Christopher Bailey for this reference. Irene Klaver, "Silent Wolves: The Howl of the Implicit," in *Wild Ideas,* ed. David Rothenberg (Minneapolis: University of Minnesota Press, 1995), 117–131. I am grateful to Dr Klaver for sending me a copy of her chapter.

36. R. H. Fuchs, *Richard Long* (London: Thames and Hudson, 1986); R. Mabey, S. Clifford, and A. King, eds., *Second Nature* (London: Jonathan Cape, 1984), 80. Land art, which clearly has both permanent and ephemeral aspects, is surveyed by J. Beardsley, *Earthworks and Beyond: Contemporary Art in the Landscape* (New York: Abbeville Press, 1984) and by James Dickinson in this volume.

37. A. Graham-Dixon, "Blinded by the Light," *Independent,* 3 December 1996, 4–5. K. Bazarov, *Landscape Painting* (London: Octopus Books, 1981), 131. C. Waite, *The Making of Landscape Photographs* (London: Collins and Brown,

1992), 40–47; Fay Godwin, *Land* (London: Heinemann, 1985). The advertiser was Ernest and Julio Gallo Winery of California, and the advertisements appeared in various issues of the *Independent Magazine* in October 1996. The ads showed sand dolphins made by Thomas Shea at Pfeffer Beach, Big Sur, California, accompanied by the following copy: "Endless time, patience and passion, yet probably gone by the evening end." The product advertised was Gallo's Turning Leaf (i.e., turning red in the fall) cabernet sauvignon. Subsequently, a leaf sculpture made by Jim Hynde was used to advertise the same wine.

38. R. S. Ulrich, "Visual Landscapes and Psychological Well-Being," *Landscape Research* 4 (1979): 17–23. S. Kaplan, "The Restorative Benefits of Nature: Toward an Integrative Framework," *Journal of Environmental Psychology* 15 (1995): 169–182; C. M. Tennessen and B. Cimprich, "Views to Nature: Effects on Attention," *Journal of Environmental Psychology* 15 (1995): 77–85.

2

Journey into Space

Interpretations of Landscape in Contemporary Art

JAMES DICKINSON

AT THE CONCLUSION of his classic study *Landscape into Art,* Kenneth Clark lamented forces at work in the mid-twentieth century threatening to dislodge landscape from its preeminence as the most popular subject of painting and the principal source of pictorial beauty. Clark feared the rise of new art styles, arguing that "the more austere forms of abstract art" could not constitute "a possible basis for landscape painting." Yet he felt scientific advances—epitomized by the atomic bomb—had so radically altered our conception of nature, as well as our ability to control it, that "the snug, sensible nature which we can see with our own eyes has ceased to satisfy our imaginations." While for Clark it was inevitable that the "excitement and awe" of this "terrible new universe . . . will find expression in some way," the precise form it might take—and thus the future role of nature in art—he claimed, "we cannot foretell."[1]

Although he revised his book as late as 1976, Clark seemed curiously unaware of a radical new landscape art that by the 1960s was working to express the "terrible new universe" he so feared. New artists were discontented with pictorial depictions of landscape and searched for ways to make art in and of the spaces that lay beyond studio, gallery, and museum.[2] This essay focuses on three distinctive interpretations of

Research for this essay has been supported by a Rider University Research Leave and a Summer Research Fellowship. Jeanne-Claude kindly provided information on projects by the Christos. I would like to thank Peggy Lewis for her excellent editing of an earlier version.

landscape advanced by the new open-space art: landscape as information, as spectacle, and as entropy associated with the work of Dennis Oppenheim, the Christos, and Robert Smithson, respectively.

Open-space art first appeared in the late 1960s and early 1970s, when artists such as Michael Heizer, Robert Smithson, Dennis Oppenheim, Robert Morris, Nancy Holt, and others began to create works for the real spaces of society, spaces either geographically or socially far beyond the privileged world of studio, gallery, and museum. Influenced by minimalism, abstraction, and conceptual art, these artists began to explore the aesthetic potential of different landscapes, developing an alternative art system that was one of the most important innovations of the American avant-garde.[3] The new landscape art rejected the art object as a commodity, as well as the gallery system that underlay commodification. It encouraged considerable innovation with nontraditional materials and production techniques to create a startling array of outdoor artworks, and it developed new modes of consuming the art object and hence new conceptions of the viewing subject. The emergence of open-space art was not only closely connected with the drive to develop a comprehensive critique of modern art's often hidden presuppositions, but also an analysis of the role of space in the constitution of the art object. This led to an investigation of the art-making potential of open spaces.

It had became increasingly apparent that the spaces within which art was produced and consumed differed greatly from the ordinary spaces where most people lived and worked and that this difference accounted for some of the distinctive characteristics of modern art produced in the studio and consumed in the gallery or museum. The artist's studio was unlike other production sites in society, generally lacking an extension of the technical division of labor, regimentation by technology, and elaborate bureaucratic structures. Using relatively simple materials such as brushes, paint, and canvas, the artist typically worked in a craft fashion, producing unique, portable works for sale while retaining complete control over the studio domain.

Moreover, the spaces of display exerted a powerful influence on the art object. As O'Doherty observed in his brilliant essay on the gallery, "the history of modern art can be correlated with changes in that (gallery) space and in the way we see it. We have reached a point where we see not the art but the space first."[4] The architecture of the gallery "imposed a way of seeing . . . that was, in many ways, even stricter than

Renaissance perspective."[5] As a careful arrangement of walls, corners, and floors, the gallery was "constructed along laws as rigorous as those for building a medieval church."[6] It created a "frame" that overwhelmingly favored two-dimensional painting and, to a less extent, pedestal sculpture.[7]

The drive to explore the aesthetic potential of real spaces was triggered by a revival of interest in the 1960s in the critique of modern art developed by Marcel Duchamp earlier in the century. For Duchamp, of course, it was a mistake to think of art exclusively in terms of aesthetically pleasing objects produced by an inspired genius in the course of some alchemical transformation of seemingly dull and ordinary materials. On the contrary, art making was essentially an intellectual process of designation and contextualization. To demonstrate his point, Duchamp invented the Readymade, an ordinary object such as a snow shovel, a bottle rack, or a bicycle wheel the artist selected and subsequently transformed into a work of art through designation and reversal, a process involving a subtle collaboration with audience. Through this extraordinary device, Duchamp hoped to undermine the elitist, retinal, and highly serious character of modern art and overcome the traditional separation of art from life.[8]

The Readymade had a number of revolutionary implications that a new generation of American artists, intent on demolishing modern art, were eager to take up. It suggested that art making was largely an intellectual, abstract process of naming things rather than a matter of individual creativity, an idea particularly appealing to the swelling ranks of arts graduates eager to make an impression on the burgeoning cultural scene.[9] The Readymade highlighted the role of space or site in the constitution of the art object. On closer examination, the Readymade suggested a critique of not only the conventional art object, but also conventional art spaces. Logically, if the art object was to be drawn from everyday life, the spaces of everyday life—the real spaces of society—were where art could and indeed should be found.[10] A number of artists enthusiastically took up the idea of fusing art making with ordinary or real space, introducing a new element that McEvilley has referred to as the "ethical imperative of site." Art had separated itself from the life world inside the sheltered environments of museums and galleries; the site was in the outside world. As a real-life object could be designated an artwork, so a real-world place could be designated an art context, and "the fact of a work of art being embedded in the real world was to be of its essence."[11] This concern with space or context allowed for the relocation of meaning from what Rosalind Krauss has called the "privacy of psychological space" to external, public, or "cultural space," thus contributing to a postmodern sensibility.[12]

Over the past three decades, artists have deployed four principal forms of open-space art, three of which are discussed here.[13] In *land projects or earthworks,* the artist directly modifies the landscape; incorporates manmade materials into the landscape; organizes complex interactions of weather, light, and landscape; or seeks aesthetic effect from aligning the landscape with celestial systems such as the sun and other stars. The art object appears as an artistically modified landscape that, because of its fundamental materiality, resists the complete reduction to documentation in media and thus suggests the continuing validity of direct audience contact with artworks. In *ephemeral landscapes,* the artist makes impermanent modifications to the landscape, which disappear either as the result of anticipated natural processes or because the artist dismantles the work. The art object, after a brief realization *in situ,* is literally dissolved as a material thing and transformed into photographs, videos, writings, and other documents. *Dialectical landscapes* are gallery installations ("nonsites" in Smithson's terminology) which present the viewer with ensembles of information about otherwise remote or undertheorized landscapes ("sites"), such as strip mines, ecological systems or decaying industrial zones.

Ephemerality, a key characteristic of the new landscape art, suggests its involvement with the dematerialization of the art object, the disappearance of the palpability that is traditionally the focus of the art making and reception.[14] Many open-space artists, including Oppenheim and the Christos, have specialized in making transitory artworks. Likewise, Smithson's works, ranging from *Partially Buried Woodshed* to *Spiral Jetty,* may be thought of as ephemeral art since Smithson intended them to "decay naturally."

In ephemeral landscape art, the artist typically emphasizes conceptual and representational aspects of the work so that these moments eventually exceed and displace the presentation of the realized work. Primarily the landscape is engaged as a "field of activity," but this is typically done to leave "no trace of this activity in the landscape." Rather, the artwork, after a brief realization, continues to exist not so much as a transformed site but as a record in words or photographs, with the documentation becoming "the only evidence of the work," which ultimately may acquire an "energy" equal to or even exceeding that of the original.[15]

As mentioned, the interpretations of landscape as information, as spectacle, and as entropy are associated with Dennis Oppenheim, the Christos, and Robert Smithson, respectively. Each has contributed to the development of open-space art, using unusual materials and techniques to explore the aesthetic properties of unconventional places. Despite these similarities, important differences exist among these artists with respect to the selection of and meaning attached to the sites chosen, the form of aesthetic intervention into the landscape, sensitivity to the issues of audience and documentation, and the role and significance of ephemerality as a constitutive aspect of the artwork.

Extending art making into open space gave rise to questions. Which of the many "real spaces" in society were suitable, or desirable, for art making? Precisely how could the artist intervene in the chosen landscape? What materials were appropriate, and on what scale would the artist now work? With the art no longer displayed at the usual locations, who comprised the audience? Oppenheim responded by making highly conceptual interventions into landscapes he described as "ravaged sites." He wanted to bring nontraditional sites such as "a scene of urban decay and desolation, or strip-mined area . . . into the expanded realm of art."[16] These sites were marginal, derelict, or otherwise unwanted areas such as chemical dumps or parking lots—places that not only lay beyond the official art world, but also were not incorporated into other spatial systems of control, such as parks and corporate plazas. With few people interested in regulating their use, the artist was free to experiment without interference. Ravaged sites avoided association with any pastoral ideal or cult of nature and signaled that moving outdoors was not part of an anti-intellectual response to the alleged "cultural sophistication" of studio-based art. As Oppenheim explained, "I was drawn to ravaged sites. When I wanted to undertake a piece, I would go to New Jersey and stomp around chemical dumps . . . Sites were places that had not been incorporated into a system—dumps, borders of countries, deserts, and wastelands—peripheries. If the land wasn't degenerate enough for me, I'd write 'diphtheria' on the hillside. The idea was a severe disjuncture from the pastoral."[17] Indeed, Oppenheim felt no obligation to play any role in repairing the appalling damage to the environment.[18]

Oppenheim did not want the new open-space art to become merely an extension of studio-based activities and ideologies, a replication, albeit on a monumental scale, of the self-referential aesthetic gestures of action painting and abstract expressionism. His landscapes of the late 1960s continued the critique of modernist art practices he had begun in his overtly Duchampian gallery-based works such as *Viewing Stations* and *Decompositions*. For example, *Directed Seeding* was a wheat field harvested so that the resulting patterns formed an "oversized parody of abstract expressionism and painterly composition." *One Hour Run* was a sort of gigantic action painting produced by running a snowmobile "intuitively and expressively" over a snow-covered landscape.[19]

However, Oppenheim also wanted to use the aesthetic exploration of real space to contribute to the new "systems art" that at the time Jack Burnham had identified as one of the most important art trends. For Burnham, modern societies are increasingly organized around largely abstract systems of knowledge and the relationships established between people and their environment. Consequently, "we are now in transition from an object-oriented to a systems-oriented culture." Thus "the specific function of modern didactic art has been to show that art does not reside in material entities, but in relations between people and between people and components of their environments."[20] Oppenheim's conceptual landscapes were designed to reference these abstract systems and relationships.

His principal innovation was to present landscape art in the form of what he called a "line of information." He would mark, divide, or otherwise manipulate the landscape in a way that allowed the modified site to reference larger fields of association beyond the art itself. The meaning of the site was, in effect, transposed from one semantic context to another. As Oppenheim put it, "When I did lines in the snow, lines which came from a map, I referred to them as information lines. They may have looked like abstract gestures, even Abstract Expressionist gestures, but the intent was to suture the work with lines or notations that had larger fields of association. Lines that could mean rainfall or temperature."[21]

Lines of information organize one of Oppenheim's most famous ephemeral landscapes, *Annual Rings* (Figure 2.1). For this work, he cut a series of semicircular rings into the winter ice that extended from both sides of the partially frozen St. John River at a point on the Maine–New Brunswick border where it formed the boundary between the United States and Canada. Neither the site itself (the river) nor the discernible object (the rings cut into the ice) had much aesthetic value in itself. For Oppenheim, significance lay not in aesthetic gesture but in the way the work referenced systems of information beyond any configuration of river and ice. The river was not just a natural phenomenon but a geo-political marker, a political boundary as well as a division in the abstract convention of time zones. The rings in the ice also suggested stylized tree-growth rings. Thus, the ensemble of "lines" that made up *Annual Rings* referenced systems of political power, information (time), and energy (biological growth) lying beyond the art itself. It was through this capacity of landscape to refer to abstract systems that Oppenheim hoped to make his contribution to systems art.

Other lines of information appear in *Contour Lines,* where Oppenheim transposed the cartographic lines on a map indicating a hillside to flat marshland periodically flooded by the tide. The wooden boards arrayed on a freeway embankment for *Landslide* referenced, in their proportional layout, cartographic lines of longitude, as well as the bleachers of a sports complex, thus introducing viewing, leisure, and (perhaps more ominously) spectacle into this otherwise marginal and functionally remote site.[22] Information lines could reference more than technical and scientific systems. For example, *Salt Flat*—a rectangle of salt spread out on a Manhattan parking lot—directly expressed Oppenheim's economic well-being since its size was determined by the amount of money he had available that day to spend on salt. For *Branded Mountain,* Oppenheim burned a large circled X on a Californian hillside, marking the landscape as ranchers brand cattle to establish ownership.

Ephemerality and transience have great importance in Oppenheim's early work, and he keenly anticipates the work's ultimate dissolution. Oppenheim creates ephemerality by deliberately "siting the art event within the agricultural and climatic cycles of nature."[23] Thus, *Annual Rings* and *One Hour Run,* to name but two works, necessarily disappeared as the ice and snow melted with the coming of spring. Likewise, *Directed Seeding, Canceled Crop,* and *Branded Mountain* slowly dissolved with the changing seasons and the onset of new cycles of growth and decay. *Whirlpool,* a vortex formed high in the air by the exhaust of a

Figure 2.1 *Annual Rings* by Dennis Oppenheim, 1968. U.S.A./Canada Boundary at Fort Kent, Maine, and Clair, New Brunswick. 150′ × 200′. Schemata of annual rings severed by political boundary. Time: U.S.A. 1:30 p.m. Time: Canada 2:30 p.m. Courtesy of the artist.

small plane, radically extended the idea of ephemerality: dissipated by air currents, it vanished even as it was created.

For Oppenheim, ephemerality had two functions. On one hand, a transient work of art undermines the modernist delusion that art is autonomous from nature and hence a gateway to immortality; the ephemerality principle thus contradicts the idea that an ahistorical or universal set of values might inform art or be embodied in it. On the other hand, ephemerality and transience provide an additional opportunity to contribute to systems art. For Burnham, the dematerialization of the art object, already apparent in the reduction of modernist sculpture to a "squiggle in the air," was itself evidence of a move toward systems art. Burnham associated this turning away in contemporary culture from an obsession with the fixed, static qualities of objects with a new interest in the science-based world of information and relationships. In aesthetic terms, this development led art to focus increasingly on "matter-energy-information exchanges and away from the invention of solid artifacts." The idea of systems art, therefore, "runs counter to the notion of the irreplaceable work of art"; it works to replace "place-oriented object art" with the "extreme mobility of systems art."[24]

Ephemerality, of course, raises questions about representation of the otherwise absent object as well as the preservation of the work of art through documentation. In contrast to the Christos' obsessive concern to document and control the images of their work, Oppenheim has been remarkably indifferent to the issue of documentation, and many of his landscapes are barely recorded in photographs. He once remarked that all he had to show for two years of intensive involvement with the landscape was just "one small box of slides." Oppenheim's reservations about documentation derived from theoretical rather than practical concerns. He held that many of his works were supposed to vanish virtually as they were being built; they were simply not supposed to go on to a "second life" as documentation.[25] He also was somewhat hostile to photography, the principal method of preserving works as an archival record. Photographs, he argued, reduced what were intended to be distinctly conceptual works to the pictorial, inevitably distorting the essential meaning and intent of the piece. Moreover, the meaning of a photographic image could change over time, further distorting the work's original purpose.

Oppenheim appears to have been indifferent to the part audience and the context played in his art. As he put it, "I have very little concern or patience for any kind of deliberation of context . . . where the show is, what the context is."[26] Again, in contrast to the Christos, Oppenheim appears to have been unconcerned to draw in the masses to activate his temporary landscapes, and, unlike Richard Serra, Robert Morris, and other land-project artists influenced by phenomenology, Oppenheim was never interested in creating works that would reconstruct the viewing subject on the basis of manipulation of visual perception and physical locomotion. As a result, Oppenheim's landscapes were not site specific in the sense that they took into account the architectural or topographical features of the site; for him, the landscape was simply a surface on which to inscribe conceptual lines of information.

Likewise, Oppenheim considered working for large audiences detrimental. "The audience doesn't make any difference. Some of the best work I've done has been done in obscure remote places without an audience . . . The supposition of a large audience upsets the process." In fact, Oppenheim finds his indifference to audience perplexing, for, as an artist, he recognizes that though he is communicating ideas, he is not sure what they really are or to whom they are being communicated: "This has been extremely disconcerting to me because I know I must be doing work to communicate . . . I know I am silently addressing an invisible audience, but I'm never aware of the audience controlling the work I do."[27]

By the early 1970s, the radicalism of the new landscape art movement was largely exhausted. Though conceptually informed works such as those championed by Oppenheim increasingly lost favor, interest in ephemeral landscape art did not disappear. On the contrary, the Christos began to undertake projects whose monumental scale and return to the grand aesthetic gesture combined to advance the idea of landscape as spectacle. Since the 1960s Christo and Jeanne-Claude have become well known for their production of enormous temporary monuments such as *Wrapped Coast, Valley Curtain, Running Fence, Surrounded Islands, The Pont Neuf Wrapped,* and *Wrapped Reichstag.* These works entailed wrapping whole buildings or other structures or otherwise marking and dividing the landscape with vast amounts of synthetic fabric. The scale and complexity of the Christos' projects require the mobilization of large amounts of labor and materials, and typically the projects take several years to plan and execute. All steps in the preparation and realization phases, including the process of gaining permission to install the works from public agencies, politicians, private landowners and other interested parties; selecting and testing the materials; commissioning planning and engineering studies; submissions to the courts and permit agencies; and so on, are regarded as part of the artwork and the art-making process.[28] Progress of the work from conceptualization to realization to removal is recorded, and the resultant documents and photographs form the basis of a comprehensive visual and written record.

The theatricality and popular appeal of these massive projects, as well as the conviction that landscape is expressive or indicative of cultural meaning, may be traced to Christo's early life experiences in socialist Bulgaria. However, Christo and Jeanne-Claude's success in realizing these enormous monuments is more a result of their mastery of the political and economic dynamic as well as of the organizational capacities of contemporary capitalism. As a student at the Fine Arts Academy in Sophia from 1953 to 1956, Christo was immersed in a world governed by the doctrines of socialist realism. His teachers at the academy familiarized him with the dimming legacy of the early Soviet avant-garde, with its advocacy of the theatricalization of space as a way to undermine the elitist tendency to separate life and art.[29] Christo devoted weekends and holidays to advancing the cause of international socialism by helping to paint and mount the giant portraits of political leaders that adorned public buildings and by traveling about the countryside to instruct peasants on how to beautify the landscape by the "orderly stacking of hay in summer and the fashioning of parapets of snow in winter."[30] One important project assigned to Christo and his comrades was the beautification of that part of the countryside through which the Orient Express passed on its weekly run through Bulgaria. The student brigades rearranged the countryside less to benefit local farmers than to transmit the appropriate ideological messages (socialist prosperity, proletarian power) to the captive observers in the passing train. Christo came to understand that even "natural" landscapes are imbued with cultural meaning that art might enhance and manipulate. He also learned that to secure the insertion of this meaning into culture takes mass populist involvement.[31]

Influences from Christo's youth are perhaps most evident in *Running Fence,* a 24.5-mile sinuous fabric curtain across the California hills that can be read as both a train moving through the landscape and a Cold War border dividing East from West (Figure 2.2). As one commentator put it, "If *Running Fence* is a parodic copy of the Berlin Wall, it is also the phantom of the Orient Express. It runs through the countryside; it weaves through the valleys like a train; and it has direction. Like the fugitive the train kept trying to glimpse, it moves from East to West."[32]

The Christos' approach to landscape differs substantially from Oppenheim's and, indeed, from land art in general. This is apparent if we look at the meaning that site has for Christo and Jeanne-Claude, their form of intervention into the landscape, their attitude toward audience and documentation, and their conception of ephemerality. The Christos most often work with sites occupied or marked by human use and activity. Several factors influence their selection of site. For wrapped building or structure projects such as *The Pont Neuf Wrapped* or *Wrapped Reichstag,* political and historical meaning are typically paramount. For other

FIGURE 2.2 *Running Fence, Sonoma and Marin Counties, California* by Christo and Jeanne-Claude, 1972–76. Height 55 meters (18 feet); length 40 kilometers (24.5 miles). Copyright 1976 by Christo. Photo by Jeanne-Claude. Courtesy of the artists.

projects, notably those involving lateral extensions in space, sites are chosen for topographical features that enhance the anticipated visual effect of introducing a grand abstract gesture into the landscape, as with *Running Fence* and *Valley Curtain,* or for their proximity to population centers, as, for example, *Surrounded Islands* and *The Umbrellas, Japan-USA* project.

For sites imbued with political or historical meaning, successful realization depends, in the first instance, on Christo and Jeanne-Claude building what they call a "political machine" that is made up of powerful individuals who control, or have the ability to control, the site and acts as a pressure group securing permission for them to proceed. For projects with a pronounced lateral extension, the Christos may have to secure cooperation from a multitude of private landowners, as well as the public agencies that regulate local land use.

Because of the enormous scale and complexity of many of the projects, Christo and Jeanne-Claude also have to build an "economic machine" capable not only of assembling the capital required, but also of planning and constructing these elaborate monuments. The Christos organize themselves as a private corporation, whose expert fund-raising and managerial capacities assure access to commercial loans usually available only to business corporations.[33] Moreover, since all projects are entirely self-financed, receiving no public money, the Christos fund each project through the sale of conventional artworks such as drawings, collages, and models.[34]

Typically, the Christos' grand aesthetic gestures are not conceptual. A project does not refer to some wider field of association (as does Oppenheim's works) but produces a beautiful object, a sensational aesthetic experience, on a temporary, monumental scale. Projects remain *in situ* for a short time, usually two weeks, after which they are removed and the materials recycled. In this regard, Christo "looks away from first things, principles, categories, supposed necessities . . . towards last things, fruits, consequences, facts." His art "focuses on the processes of work and communication rather than on concerns of quality and form."[35]

The Christos work exclusively with industrially produced synthetic fabric that is altered and dyed to exact specifications. Fabric color is extremely important and accounts for a good portion of the overall aesthetic impact of any project. Moreover, by using synthetic materials and insisting on their recycling at the conclusion of each project, the Christos can reference the ultimate capitalist commodity—oil and its derivative products—and thus tie the aesthetic dimension to the material basis of contemporary civilization.

The Christos are extremely sensitive to context and audience. Since they need support from elites to ensure the realization of projects, they must spend time and energy soliciting it. In *The Pont Neuf Wrapped* book, Christo and Jeanne-Claude are shown in meetings with leading French politicians, including the prime minister of France and the mayor of Paris. Getting permission to wrap the Reichstag (a twenty-four-year process concluded in 1995) depended on support groups the Christos formed in Germany that comprised leaders in politics, business, science, and the arts who could argue the project's merits in the halls of power, in meetings documented in great detail in the book of *Wrapped Reichstag*. The Christos are necessarily very concerned with documentation. As a project unfolds, all aspects of conceptualization, planning, and execution are obsessively recorded and later reproduced in massive books, photographic presentations, museum displays, lectures, and videos. Not only is documentation extensive, but the Christos' organization rigorously controls the production and dissemination of images. The paper trail accumulated in the course of satisfying technical, legal, and environmental requirements is included as part of the work. Documentation thus far exceeds the gallery or installation photograph that usually records an artwork. For the Christos, the art object as media transcends the gallery. This record becomes, with the disappearance of the original work, the only available (but nonetheless secondhand) version of the altered landscape.

The Christos depend on mass public participation to "activate" their works, exposing the "transient beauty of the ephemeral" of the work *in situ.* Although they regard this moment of the work's existence as primary, popular appeal turns the temporary monument into event and experience, and consequently the impermanence of the material work passes into the "permanence" of cultural memory. The grander the project, the more essential it becomes to maximize public participation, and the projects walk an increasingly fine line between aesthetic experience and spectacle.

Finally, the Christos manipulate ephemerality, not as Oppenheim did to demonstrate that art is grounded in other systems, but as a cunning way to secure immortality for both the artwork and the artists. As Laporte has put it, the Christos use impermanence and transience to "create a work of memory that subverts our arithmetical and linear image of eternity, so dependent on its representation in stone."[36] Consequently, their landscapes do not explore the alleged permanence of materials. They aim to render the artwork as memory. This goal is based on the project's ability to stimulate in the viewer the "involuntary beauty of the ephemeral"—an "untimely hallucination that upsets the order of the world for a moment and then vanishes." The landscape ultimately returns to its former familiarity, but it remains, along with the audience, "definitely haunted by the intrusion."[37] The "involuntary beauty of the ephemeral" is enhanced not only by the viewer's anticipation of the disappearance of the work, but also by the aesthetic properties of its form and materials: the undulating fabric, the subdued earthlike tones and colors, and the wrapping that washes out distinctions and particularities of site. Thus transformed from monument into event, the work of art translates into experience and finally, with the help of elaborate documentation, into memory.

Landscape as entropy, or dialectical landscape, constitutes a third interpretation of landscape. Like other forms of open-space art, landscape as entropy deals with the relationship of presence and absence, experience and memory, and the real and the ideal, but it tackles these issues in a highly original way: as a dialectic of entropic "site" (actual landscape) and "nonsite" (gallery display of this landscape). With ephemeral landscape, the realized art object, whether modified site or landscape, dissolves in favor of documentation and representation in mass media. The photograph, slide, book, or essay supercedes the traditional locus of display, the gallery. Moreover, little is gained by returning to the place the work was originally realized or displayed, say, to Miami Bay to gaze on the site of *Surrounded Islands* or to the hillsides of California to reimagine *Running Fence.* What is important is not so much the presence or trace of the object in the landscape, or the landscape itself, but rather the trace of the object recorded in various media. In contrast, with land projects or earthworks, the art object retains a strong presence as a material thing that modifies an actual landscape, and it resists reduction to a representation. The artwork has a double existence: as real object (site-specific work of art) and as ideal documentation (representation of that work of art). That a land project tends to be known principally in its mediated, that is, documentary, form rather than directly as a physical and visual object is largely the (accidental) result of the work's physical remoteness. Knowing the work indirectly, as media, does not correspond to the work's essential property. A visit to an earthwork, to observe and walk around the object offers (in theory at least) a superior mode of appropriation to that gained by any media.[38]

Dialectical landscape does not modify the landscape physically, either temporarily or permanently. It is conceived of as the site of powerful, abstract forces that the artist seeks to bring to consciousness (and hence intellectual awareness and control) by means of a complex documentary representation, typically in an elaborate gallery installation. In this way, the artist captures a real-world landscape, or *site,* in an installation called a *nonsite.* Yet at the moment this appropriation of an absent landscape is achieved, its representation, the nonsite, becomes inadequate, stimulating the viewer's return, both intellectual and physical, to the actual landscape itself, the site that informs the nonsite.

The work of Robert Smithson is central to the development of this dialectical landscape concept. In his brief career (he was tragically killed in 1973 in a plane crash while surveying terrain for an earthwork project in Texas) Smithson made major contributions to gallery-based minimalism (*Glass Stratum, Plunge,* and *Alogan*) as well as to land projects and earthworks (*Spiral Jetty* and *Amarillo Ramp*). Although a pioneer in exploring the artistic potential of open space, he was virtually alone among his contemporaries in remaining interested in the role of galleries and museums in displaying and defining the postmodern art object. His major contributions were both to conceive of the landscape as a site embodying the immense and powerful forces of entropy and to devise a method of appropriating this landscape as a dialectic of site and nonsite. This derived from his novel understanding of space in art and his appreciation of the postmodern landscape as the new sublime.

According to Hobbs, Smithson thought that sculpture (broadly conceived) should concern itself with the absence of space, that is with the "voids that displace the solidity of space." The most interesting voids for Smithson seemed to be those open spaces typically "assimilated only in areas society has abandoned, its waste areas, unrecognized danger spots, excavations." Sculpture as void thus concerns itself with what can best be understood as nonspace, those "immediate surroundings that fail to impinge themselves on modern consciousness." These include undertheorized spaces such as suburban tract housing, shopping malls, and superhighways, as well as more familiar but taken-for-granted spaces like the modern gallery or even the cinema. These newly recognized spaces, in their lack of intellectual presence, are what Hobbs calls "cancelled places," examples of an "ever-present nowhere."[39]

As Smithson moved away from gallery-based minimalism, he became increasingly interested in a class of sites that might be called entropic "voids" in the landscape, "places where energy has been drawn out."[40] Of course, entropy is a fundamental natural process indicative of the winding down of natural systems, including the universe itself.[41] For Smithson, however, entropy was also a characteristic of contemporary economic and social systems. Indeed, Smithson was attracted not so much to natural wonders like the Grand Canyon but to those mundane, manmade places that revealed the essentially entropic character of contemporary civilization: strip mines, tailing ponds, slag heaps, quarries, and waste dumps, as well as zones of deindustrialization formed by decaying towns and industrial rust belts. For Smithson, these unrecognized landscapes were extremely foreboding, for it was precisely here that the vast unfathomable forces of entropy, both natural and social, silently worked to dissolve the landscape, cancel the present, render experience as memory. The open spaces of contemporary society, the spaces artists were going to have to deal with if they left the gallery, were "deeply mired in entropic forces." They were "open, unfocused," and not as yet brought to consciousness.[42] A new generation of artists could address these undertheorized places.

Smithson's most memorable works use such places to explore entropy directly, not only in nature but also in the failure of capitalist technology and economic institutions. For example, he intended his iconic earthwork, *Partially Buried Woodshed,* to demonstrate the connection between entropy and the production of historical meaning: he piled earth on the structure's central beam until it cracked, initiating a process of decay in which the work "would gain in legend as it diminished in existence," reinforcing Heizer's observation that "as the physical deteriorates, the abstract proliferates, exchanging points of view."[43] His ability to uncover new meaning for wornout and mundane places is also powerfully evident in his writings, especially his classic art/essay *A Tour of the Monuments of Passaic, New Jersey.* Here Smithson turned a visit to a derelict industrial riverfront into an analysis of the ever-present forces of entropic dissolution, describing the child's sandbox in the playground as the graveyard of nature and the used-car lot as a "lower stage of futurity."[44] Entropy is also the subject of many of his photographs of industrial sites, marginal zones, and natural areas in the United States, Germany, and elsewhere and of his meditation on the endless making and remaking of the Hotel Palenque in Mexico.[45]

As the site of entropy, the landscape was for Smithson the locus of the postmodern sublime. The sublime can be understood as that "absolutely immense object or absolutely powerful event" where "the visible object dissolves and the mind is thrown back on itself."[46] But where modernism had appreciated the sublime as either nature (e.g., waves and mountains), otherness (e.g., the exotic East), or technology (e.g., the factory, the electric cityscape, or the atom bomb),[47] for Smithson, the postmodern sublime consisted of a fundamental awareness that the vast processes of universal dissolution and decay revealed in the landscape were not simply the work of nature, but "something we have made; it is us." Environmental destruction, pollution, deindustrialization, and failed technology comprised the evidence of entropic forces at work in the postmodern landscape, which "neutralizes the myth of progress," creating "a mood of vast immobility."[48] Thus industrial capitalism reveals itself as a vast entropic process that cannot, despite its claims, sustain growth and progress. Smithson's discovery of landscape as entropy marks the closure of optimistic, expansionist capitalism, a cancellation of that peculiarly American version of the sublime in which technology "transformed the individual's experience of immensity and awe into a belief in national greatness."[49]

How can art capture, frame, and cognitively contain this entropy? For Smithson, the conventional techniques used in representational painting and sculpture cannot depict such sites realistically. Thus, when confronting the sublime, "the artist can never depend on painting, but must opt for poetry"; forced to retreat from the purely visual, the artist resorts to "words."[50] Drawing on this insight, Smithson held that the entropic forces immanent within any landscape can be revealed, checked, and contained only by the accumulation and transmission of appropriate "information" about that landscape. Because information is grounded in language, it is, in fact, a form of order—a momentary respite from entropy. Language therefore constitutes the "universal center" from which the assault on entropy can be organized; it is a limitless system that can be "stretched" to form the infinite museum now needed to grasp, display, and contain the entropic site. Thus, for Smithson, artistic activity consists less of intervening in the landscape than of creating a parallel landscape, a nonsite or elaborate information package comprising three-dimensional map objects, crystals, bins of rock, and other natural materials, drawings, photographs, and written texts (Figure 2.3). The nonsite thus constitutes a documentary representation of an actual site, an ensemble of information about an otherwise absent and ungraspable place. Since the nonsite takes the form of a gallery installation, conventional art spaces retain a crucial role in the postmodern art process (Figure 2.4).

For Smithson, however, site and nonsite exist in a dialectical relationship. The actual site is, of course, neither abolished nor transcended; nor are its entropic forces contained by the nonsite gallery display. Similarly, the nonsite, however elaborate, remains an inadequate representation of the site to which it refers. Yet the two "realities" are connected

FIGURE 2.3 *A Non-Site, Franklin, New Jersey* by Robert Smithson, 1968. Wood, limestone. Courtesy of the John Weber Gallery, New York.

IGURE 2.4 *A Non-Site, Franklin, New Jersey* by Robert Smithson, 1968. Aerial map. ourtesy of the John Weber Gallery, New York.

TABLE 2.1 Smithson's Distinctions between Sites and Nonsites[51]

Site	Nonsite
Open limits	Closed limits
A series of points	An array of matter
Outer coordinates	Inner coordinates
Subtraction	Addition
Indeterminate certainty	Determinate uncertainty
Scattered information	Contained information
Reflection	Mirror
Edge	Center
Some place (physical)	No place (abstract)
Many	One

in a way that allows them to access each other. This is apparent if we look more closely at how Smithson distinguishes site from nonsite in terms of two sets of opposed but related terms, shown in Table 2.1. These mental or categorical distinctions duplicate the physical separation of site from nonsite: the distinct, bounded, and contained space of the nonsite contrasts with the open, unbounded space of the site. Yet this separation sets up a complex relationship that links the two dialectically by creating, as Smithson put it, "a course of hazards, a double path made up of signs, photographs, and maps that belong to both sides at once." He continues:

> Both sides are present and absent at the same time. The land or ground from the Site is placed *in* the art (Nonsite) rather than the art placed *on* the ground. The Nonsite is a container within another container—the room. The plot or yard outside is yet another container. The two-dimensional and three-dimensional things trade places with each other in the range of convergence. Large scale becomes small. Small scale becomes large. A point on a map expands to the size of the land mass. A land mass contracts to a point . . . Is the Site a reflection of the Nonsite (mirror), or is it the other way around?[52]

In the gallery installation, the nonsite becomes the signifier for the absent site, but it also, as a complex documentary representation, begins to take on the characteristics of a "real" place. The gallery installation becomes a new site, and the separation of the site and its "containment" in the gallery has broken down. The two realities become fluid and merge; the viewer can pass from one reality to the other and back. Rules that govern movement back and forth on this "double path" of signs "are discovered as you go along uncertain trails both mental and physical."[53] Because they are inherently inadequate representations of real sites, according to Hobbs, "the non-sites exude a feeling of uneasy containment" and the persistent and unsettling "feeling that something is missing" can be offset only by the gallery visitor's gaining greater knowledge of the original site. Thus Smithson's dialectic of site and nonsite propels the viewing subject into physical exploration and mental appreciation of the landscape. As Hobbs concludes, "Smithson wanted to show that art in the gallery is a diminution of a far more interesting activity—going to actual locations." Smithson himself would agree, suggesting that moon rocks on display in the museum induce a powerful desire for space flight! The result of Smithson's dialectic is an intellectual restlessness that undermines the sanctity of the art object and propels the viewer into physical locomotion and exploration of the real world. "Acting in a dialectic with the sites, the nonsites negate both the gallery's space and the primacy of perception as they point away from themselves to the site. In this way the art object's actual presence is undermined."[54]

Yet exposure to the nonsite modifies how the actual site will be experienced. Without the nonsite, the real-world landscape may remain hidden. A comparison between the landscape rendered as nonsite and its direct experience as site allows assessment of the nature and degree of change within the landscape, a "measurement of entropic forces." By moving back and forth between site and nonsite, the real and the abstract, experience and memory, the viewing subject can become aware of, as well as contain and control, those entropic forces continually threatening the dissolution of the object and hence of the subject.

In the dialectical form of open-space art, landscape is thus rendered as the immanent information it contains, and this in turn is displayed and contained in the gallery as art, as nonsite. Here Smithson shows his sensitivity to issues of presentation and audience within a postmodern art system and his reluctance to abandon the gallery format altogether. Indeed, for Smithson, it is precisely in the controlled world of the gallery that entropy can be raised to consciousness, the absent landscape made present, and the attendant chaos and disorder of the real world incorporated into human experience.

Open-space art provides an important but little-recognized window

Table 2.2 Three Interpretations of Landscape

	Oppenheim	Christo and Jeanne-Claude	Smithson
FORM OF LANDSCAPE INTERVENTION	artwork as "line of information" on landscape; references larger fields of association, e.g., systems of science, technology, politics lying beyond art; rejection of sudio ideologies; conceptual character related to systems theory of Burnham	artwork as grand aesthetic gesture on the landscape; synthetic fabric used to wrap buildings & structures, or mark & divide the land; monumental scale with elements of spectacle; has no larger field of reference; extension of studio ideologies into landscape	artwork as investigation of entropy either by direct intervention into landscape, e.g., *Spiral Jetty, Partially Buried Woodshed;* or indirectly, by appropriation of site as nonsite, an ensemble of information about an absent, unmodified site or landscape
SITE	chooses "ravaged sites"—waste dumps, derelict land, or marginal, boundary sites demarcating divisions within and between systems; rejects pastoral ideal and cult of nature; no obligation of art to repair environmental damage	sites/landscapes typically marked by human usage; chosen for political/historical meaning, e.g., *Wrapped Reichstag,* desired topographical features, e.g., *Running Fence;* or public accessibility, e.g., *Surrounded Islands.*	voids, canceled places, entropic zones; undertheorized, marginal landscapes and places which embody natural and social forces of change and entropy; gallery installation become nonsite
AUDIENCE/ CONTEXT	insensitive to receiver-end of art process; indifferent to audience and resistant to context; avoids commissions for site-specific works; produces work without clear idea of what is to be communicated or to whom; anti-spectacle	very sensitive to receiver-end of art process; developed sense of context and audience; projects require approval from interested parties controlling landscape; masses "activate" temporary monument as event and experience	dialectic of site/nonsite involves audience in mental and physical movement between installation and real world place; gallery thus still plays crucial role in presentation and definition of art object
DOCUMENTATION	minimal; resistant to photographic recording because: (i) distorts conceptual nature of work by reduction to the pictorial; (ii) meaning of images changes over time; works meant to disappear	extensive and tightly controlled; detailed record of conceptualization, planning, execution and *in situ* presentation of work in books, videos and exhibitions	extensive documentation creating multiple versions of work; work exists in many forms: as real object, photo, essay, video which may have power and imagination equal to, or exceeding, the original
EPHEMERALITY	work disappears as it is produced, or as result of natural processes and cycles; functions to ground art in other systems, e.g., nature; and critique conventional ideas about art object and transcendent values; dematerialization reflects connection to systems art	work deliberately removed after set time period; "involuntary beauty of the ephemeral" offers new route to immortality of art by turning monument into experience and culture memory; work thus "survives" as long as people remember and culture exists	work may decay and disappear as result of erosion, entropy; moves from real to abstract, acquiring meaning as it disappears; comparison of site with nonsite allows for measurement of entropic change

on the changing relationship of landscape and technology. As these case studies show, open-space art provides interpretations of landscape that go far beyond that possible in conventional art such as painting. By interacting with a variety of real-world landscapes, these interpretations offer a fascinating look at not only aesthetic innovation but also the meaning of landscape and its relation to technology and change in contemporary culture. As a result, landscape now occupies an even more prominent place in contemporary art and cultural discourse than Kenneth Clark envisioned.

NOTES

1. Kenneth Clark, *Landscape into Art,* rev. ed. (London: J. Murray, 1976), 237–241. Of course, many others have investigated the social, economic, and political transformations underlying the emergence and ascendancy of landscape painting and pointed to the deep ideological significance of landscape in modern culture as well. See Ann Bermingham, *Landscape and Ideology* (Berkeley: University of California Press, 1986); John Barrell, *The Dark Side of the Landscape* (Cambridge: Cambridge University Press, 1980); and Simon Schama, *Landscape and Memory* (New York: Knopf, 1995).

2. John Beardsley, *Probing the Earth: Contemporary Land Projects* (Washington, D.C.: Smithsonian Institution Press, 1977), 9.

3. For general discussions of the new landscape art, see Rosalind E. Krauss, *Passages in Modern Sculpture* (Cambridge, Mass.: MIT Press, 1977), esp. chap. 7, and Henry M. Sayre, *The Object of Performance* (Chicago, Ill.: University of Chicago Press, 1989), chap. 6. For examples of the new outdoor art, see John Beardsley, *Earthworks and Beyond: Contemporary Art in the Landscape* (New York: Abbeville Press, 1984), Gilles A. Tiberghien, *Land Art* (New York: Princeton Architectural Press, 1995), and Jeffrey Kastner, ed., *Land and Environmental Art* (London: Phaidon, 1998).

4. Brian O'Doherty, *Inside the White Cube: The Ideology of the Gallery Space* (Santa Monica, Calif.: Lapis Press, 1986), 14.

5. Sayre, *The Object of Performance,* 211.

6. O'Doherty, *Inside the White Cube,* 15.

7. An illusion-based aesthetic much criticized by Donald Judd in his seminal "Specific Objects" (1965) in *Complete Writings* (Halifax, Nova Scotia: The Press of the Nova Scotia College of Art and Design, 1975).

8. For an account of the dissemination of Duchamp's ideas in the United States, see Robert Pincus, " 'Quality Material' . . . Duchamp Disseminated in the Sixties and Seventies," in Bonnie Clearwater, ed., *West Coast Duchamp* (Miami, Fla.: Grassfield Press, 1991). For general discussions of Duchamp's Readymade, see Krauss, *Passages in Modern Sculpture,* chap. 3; J. Masheck, ed., *Marcel*

Duchamp (Englewood Cliffs, N.J.: Prentice Hall, 1975); and Calvin Tompkins, *The Bride and the Bachelors* (New York: Penguin Books, 1978).

9. In 1965 Richard Pettibone produced a series of three Readymades that directly referenced those of Duchamp. See Pincus, " 'Quality Material,' " 90.

10. Thomas McEvilley, "The Rightness of Wrongness: Modernism and Its Alter-Ego in the Work of Dennis Oppenheim," in Alanna Heiss, *ed. Dennis Oppenheim: Selected Works 1967–1990* (New York: Abrams, 1992), 7–10. Duchamp critiqued conventional gallery space in several pioneering installations, such as *1,200 Bags of Coal* (1938), *Mile of String* (1941), and *Boite-en-Valise* (1941), but ultimately he remained too rooted in "museology" to achieve as radical a breakthrough with respect to art space as he had achieved with the art object.

11. McEvilley, "The Rightness of Wrongness," 7, 10.

12. Krauss, *Passages in Modern Sculpture,* 270. As Frederic Jameson has noted, "a certain spatial turn has often seemed to offer one of the more productive ways of distinguishing postmodernism from modernism proper." Conceptual art undermines confidence in the validity of perceptual experience, since the categories of mind it seeks to uncover "can never become visible objects in their own right." We therefore can no longer find solace in the traditionally conceived "solid" work of art and must learn that "the spatial field is the only element in which we move and the only 'certainty' of an experience." Jameson, *Postmodernism* (Durham, N.C.: Duke University Press, 1991), 154–157.

13. This typology is developed in more detail in my paper "In Its Place: Site and Meaning in Richard Serra's Public Sculpture," in Andrew Light and Jonathan M. Smith, eds., *Philosophy and Geography III: Philosophies of Place* (Lanham, MA: Rowman and Littlefield, 1998), 45–72. The category not taken up here for lack of space is urban site-specific public sculpture, which introduces into the highly complex, socially charged environment of the city public artworks inspired by the abstract form, monumental scale, and phenomenological intent of remote land projects. Often, artists working in this way aim to use art to reconstitute both the aesthetics of public space and the viewer's phenomenological and political assumptions. Robert Morris's early writings on sculpture, reprinted in *Continuous Project Altered Daily: The Writings of Robert Morris* (Cambridge, Mass.: MIT Press, 1993), discuss the phenomenological consequences of large-scale, outdoor works of art. By temporally stretching the consumption process beyond the retinal glance typical of museum viewing and requiring physical locomotion, open-space art promises a revitalization of the viewing subject.

14. Lucy R. Lippard, ed., *Six Years: The Dematerialization of the Art Object* (New York: Praeger, 1973), collects an important series of documents on this aspect of the American avant-garde.

15. Beardsley, *Probing the Earth,* 9. Michael Heizer's *Hydrate*—eight boards placed in a desert gully—lasted two weeks before being swept away by a flash flood. His *Five Conic Displacements* is expected to last a little longer, as it will take about fifty years for these excavated depressions to fill up with colloidal matter, thus restoring the surface of the desert and obliterating all evidence of

artistic intervention in the landscape. Michael Heizer, "The Art of Michael Heizer," *Artforum* 8 (1969): 31–39. But even for imposing earthworks, it seems that the rate of erosion is much higher than originally anticipated, raising important conservation issues for these open-space artworks. See Marina Isola, "Monumental Art, But the Wind and Rain Care," *New York Times,* 24 November 1996.

16. McEvilley, "The Rightness of Wrongness," 16.

17. "An Interview with Dennis Oppenheim," in Heiss, *Dennis Oppenheim: Selected Works,* 138.

18. Developing a land reclamation project for the Bingham copper mine in Utah, Smithson wrote, "A dialectic between mining and land reclamation must be developed. Such devastated places as strip mines could be recycled in terms of earth art . . . Art can become a physical resource that mediates between the ecologist and the industrialist." Nancy Holt, ed., *The Writings of Robert Smithson* (New York: New York University Press, 1979), 220. Both Heizer and Morris have realized earthworks that repair the ravages of industrial land use. For an inventory of reclamation art sites, see Hilary Anne Frost-Kumpf, "Reclamation Art: Restoring and Commemorating Blighted Landscapes," unpublished manuscript, Department of Geography, Pennsylvania State University, State College, Pa., 1996.

19. With *Gallery Transplant—Cornell* nature substituted for human creativity as birds made abstract, though random, impressions in the snow. McEvilley, "The Rightness of Wrongness," 17–20.

20. Jack Burnham, *Beyond Modern Sculpture* (New York: Braziller, 1968), 30.

21. "Interview with Dennis Oppenheim," 138.

22. For a literary treatment of such ravaged sites, see J. G. Ballard's novel *Concrete Island* (New York: Farrar, Straus, and Giroux, 1974) where the entire action takes place within a derelict zone formed by a complex of highway overpasses, feeder ramps, and tunnels.

23. McEvilley, "The Rightness of Wrongness," 20.

24. Burnham, *Beyond Modern Sculpture,* 364–365.

25. "Interview with Dennis Oppenheim," 139, 144.

26. Ibid., 156.

27. Ibid., 151–154.

28. For the Christos' working methods, see Jon van der Marck, "Christo: Making of an Artist" in Christo, *Christo: Collection on Loan from the Rothschild Bank* (La Jolla, Calif.: La Jolla Museum of Contemporary Art, 1981), and Calvin Tompkins, "Christo's Public Art" in Christo, *Christo's Running Fence,* (New York: Abrams, 1978).

29. Brandon Taylor, *Art and Literature under the Bolsheviks,* vol. 1 (London: Pluto Press, 1991); van der Marck, "Christo: Making of an Artist."

30. van der Marck, "Christo: Making of an Artist," 53. See also Jacob Baal-Teshuva, *Christo and Jeanne-Claude* (Cologne, Germany: Taschen Verlag, 1995).

31. Tompkins, "Christo's Public Art," 19.

32. Dominique G. Laporte, *Christo* (New York: Random House, 1986), 46–48.

33. Marck, "Christo: The Making of an Artist," 91.

34. Total project costs range from $850,000 for *Valley Curtain* (1972) to $3,250,000 for *Running Fence* (1976) and $5 million to $7 million for *Wrapped Reichstag* (1995). Based on information kindly supplied by Jeanne-Claude.

35. Marck, "Christo: The Making of an Artist," 88.

36. Laporte, *Christo,* 36.

37. Ibid., 70.

38. Sayre argues that since in Smithson's case the distinction between concept and its realization is often blurred, his documentation of his own works in photographs, films, drawings, and writings has "become a body of work in its own right, in some ways more substantial now . . . than the sculptures and nonsites and earthworks themselves." Sayre, *Object of Performance,* 215.

39. Robert Hobbs, "Smithson's Unresolvable Dialectics," in *Robert Smithson: Sculpture* (Ithaca, N.Y.: Cornell University Press, 1981), 24–25.

40. Sayre, *Object of Performance,* 218.

41. For general discussions of entropy, see my "Entropic Zones: Buildings and Structures of the Contemporary City," *Capitalism, Nature, Socialism* 7 (September 1996): 81–95, and Robert A. Sobieszek, *Robert Smithson: Photoworks* (Los Angeles, Calif.: Los Angeles County Museum of Art/University of New Mexico Press, 1993).

42. Sayre, *Object of Performance,* 226–229. See also Gary Shapiro, *Earthwards: Robert Smithson and Art after Babel* (Berkeley: University of California Press, 1995), for an extensive discussion of entropy in Smithson's art and writings.

43. Dorothy Shinn, *Robert Smithson's "Partially Buried Woodshed"* (Kent, Ohio: Kent State University School of Art Gallery, N.D.), 2. Heizer, "The Art of Michael Heizer," 31.

44. Holt, *The Writings of Robert Smithson,* 52–57.

45. Many of these photographs are reproduced in Sobieszek, *Robert Smithson: Photoworks.*

46. Sayre, *Object of Performance,* 220.

47. For a detailed discussion of modern meanings of the sublime, see David E. Nye, *American Technological Sublime* (Cambridge, Mass.: MIT Press, 1994), chap. 1.

48. Sayre, *Object of Performance,* 216–220.

49. Nye, *American Technological Sublime,* 43.

50. Sayre, *Object of Performance,* 220.

51. Holt, *The Writings of Robert Smithson,* 115.

52. Ibid., 115.

53. Ibid.

54. Hobbs, "Smithson's Unresolvable Dialectics," 29, 22.

INVENTING LANDSCAPES

3

Abandoning Paradise

The Western Pictorial Paradigm Shift around 1420

JACOB WAMBERG

TAKE THIS verbal Rorschach test: When you hear the word "landscape," what sort of terrain immediately appears before your inner eye? My guess is that most modern Western people visualize something like the picture typically found on certain supermarket food wrappings. That is, a basically green and grassy terrain of mildly curving hills, interspersed with meadows, living hedges, grain fields, and some not too obtrusive roads and houses, all of it unfurling under a blue and sunny sky. Or, to speak in terms appropriate to this volume, a landscape in which there are technological modifications of nature but the modifications have been "naturalized," so as to create a harmony between nature and culture.

If this type of terrain has attained the status of a cliché, it should come as no surprise. In a culture that until recent decades has had its economic focus in agriculture, isn't this terrain simply what one tends to see when walking outside the cities? Furthermore, this landscape emerges merely as a banalization of a terrain that appears with surprising consistency in the high art of recent centuries, a terrain that is more or less cultivated, divided into sections, and controlled—think of Constable, of seventeenth-century Dutch art, or of Quattrocento Italian art. While landscape settings in this tradition might also include, or be ex-

---

This essay introduces, in condensed form, certain parts of my yet unfinished habilitation thesis, "Landscape as World Picture. Nature Depiction and Cultural Evolution in the West, from the Cave Paintings to Early Modernity."

clusively dedicated to, an untouched wilderness, cultivation always exists, at least as a possibility.

With my point of departure in pictorial art, I should like to call into question the simplicity of visualizing the cultivated terrain. While it is true that cultivated landscapes form a long and persistent tradition in Western art, it is even more remarkable that before a precise historical point—the 1420s—such settings simply disappear. If one takes a macrohistorical overview of surviving Western images made from 3000 B.C. to 1420 A.D., one will observe that with the noteworthy exception of ancient Egypt, the only type of cultivation characterizing the terrains of two-dimensional images—paintings, frescoes, book illuminations, mosaics, and reliefs—is enclosed units such as tree shrines or gardens. Beyond the boundaries cutting off such units from the broader stretches of nature, one will basically encounter an uncultivated wilderness, a rocky desert without any of the later signs of controlled land use such as grain fields, fences, hedges, roads, and bridges.

What could be the reason for the sudden occurrence of a technologically modified nature in painting after 1420 and its just as striking absence before that time? First, one ought to note that cultivated nature is by no means the only novelty marking Western painting after 1420. It is a subcharacteristic of what amounts to a whole pictorial paradigm shift that took place at that time—a paradigm shift that also included the emergence of perspective space; cast light; and the rendering of time, most notably in the form of weather, the seasons, and diurnal cycles. In this sense, one cannot expect to isolate some specific cause determining only the depiction of cultivated landscapes, but rather must look for large-scale forces that governed, *among other things,* the attitude toward cultivated nature.

In this essay, I make use of a macrohistorical model that is on the one hand evolutionary, in presupposing a transformation of large parts of culture in a certain, describable direction, and on the other hand structurally oriented, in assuming structural affinities between the cultural domains thus transformed. To be more specific, a structure that could illuminate the pictorial paradigm shift that occurred around 1420, and not least of all its significance in relation to technological modification, is the structure of the myth that *par excellence* concerns the institution of technology: the myth of paradise and the fall. The narrative course of this myth can be shown to be structurally equivalent to the transformation that the pictorial landscape paradigm undergoes from the Middle Ages to modernity, in the sense that in both the paradise narrative and the landscape paradigm, one moves from a state founded on a spatially restricted and timeless virgin earth to a state in which the ground is wide, subject to weather, and cultivated.

If one resorts to a sociological interpretation of the technological aspect of this dual development, an apposite key concept is the activity lying behind technology, namely, *work*. In antiquity and partly into the Middle Ages, physical work was considered a debased activity that distracted the mind from spiritual insight and whose traces therefore were seen as alien to pictorial nature. However, during the Middle Ages work was gradually transformed into first a penitential activity and later, with the advent of capitalism, a necessary duty, a common calling for everyone, regardless of social level. Instead of pointing toward a debased drudgery, grain fields now became morally uplifting, if not idyllic, and consequently also became a respectable part of the pictorial repertoire.

Before delving deeper into the argument, it is worth stressing on the one hand the suddenness with which the new landscape paradigm makes its breakthrough around 1420 and, on the other, how relatively stable over long periods both it and its predecessor were, a fact that encourages a structuralist analysis. As is seen, for instance, in a work like Lorenzo Monaco's *Adoration of the Magi* (1410; Florence, Uffizi), the old landscape paradigm requiring a rocky wilderness, a relatively shallow spatial setting, and a nonatmospheric sky (here a golden ground) still predominates in the beginning of the fifteenth century (Figure 3.1). However, when we move just two decades forward to Gentile da Fabriano's *Flight into Egypt* (1423; Florence, Uffizi) and Robert Campin's *Nativity* (c. 1425; Dijon, Musée de la Ville), the new paradigm is fully manifest south as well as north of the Alps (Figure 3.2). Now, in both scenes, the gaze is led along rural roads winding among plowed fields, while golden beams are cast from a sun behind faraway hills or mountains. Not until Picasso's cubist experiments and Kandinsky's abstractions at the beginning of the twentieth century is this type of landscape seriously challenged. If it is this type that somehow incarnates our idea of "landscape," it is no wonder, for the Germanic forerunner of the very word *landscape* (*Landschaft*) emerged only in the fifteenth century, simultaneously with the new pictorial paradigm.[1]

FIGURE 3.1 *Adoration of the Magi* by Lorenzo Monaco, 1410. Panel. Courtesy of Galleria degli Uffizi, Florence. Photograph by Fratelli Alinari.

On the other hand, in Western images harking all the way back to Minoan and Sumerian times, there are generally no changing and atmospheric skies, no far-ranging vistas, and—especially relevant to the present discussion—no cultivated grounds. That is as already mentioned, there might be enclosed cultivated spaces, such as gardens and holy precincts, but they do not expand to large stretches of earth—earth that, when exposed, is always made up of mountains and rocks. One might mention Egypt as an exception to this rule, because Egyptian art teems with depictions of rural labor and therefore also with images of grain fields. Yet, since Egyptian pictorial space never expands to vistas outside a shallow middle ground, it remains problematic whether one could talk of Egyptian landscape imagery at all.

Indeed, an argument could be made that the concept "landscape"

FIGURE 3.2 *Nativity* by Robert Campin (Master of Flémalle), c. 1425. Panel. Courtesy of Musée des Beaux-Arts, Dijon.

by definition subsumes a certain measure of panoramic surroundings, which, semiotically translated, amount to areas beyond the radius of the iconographic subject matter. Such an iconographically indeterminate space becomes possible only after land formations are suitable for serving as proper backdrops for figures, a condition realized precisely in the non-Egyptian cultures having rocks as their pictorial grounds. Basically, the lack of cultivated terrains thus belongs to these backdrops and not to the iconographic themes that potentially (for example, in a medieval depiction of the Cain and Abel story) might evoke a grain field or two. Another way of describing the innovation of landscape portrayals after 1420 is to say that cultivation enters the general spatial repertoire *independently* of what action or theme is being depicted.

Making general statements on such huge analytical units as these—chronologically, geographically, socially—is, of course, risky business, already at the empirical stage. How reliable are the surviving images in offering an impression of the visual culture of their time? Are they representative of society as a whole or only of certain groups? Can we even define a category as "visual art" and follow it through several centuries despite radically changing functions? Although these questions deserve a lengthy discussion, I must here resort to two somewhat crude postulates: (1) The images I discuss here tend to reflect the ideals of the ruling classes, and (2) in this respect they are indeed representative and not distorted by the accidents of time.

When trying to find explanations for these macrohistorical observations, an easy framework would be a purely perceptorial argument a la Gombrich.[2] Before the fifteenth century, this argument would posit, mimicry was not well developed, and image makers were restricted to simple conventional schemata, for example, mountains without cultural traces. Afterwards realism had its breakthrough, opening artists' eyes to the factual, that is, culturally worked, surroundings. Another explanatory framework of a more phenomenological, material-sensing kind would commence not in the eye but in the earthy ground: if premodern terrains consist of rocky stone, they must consequently be unreceptive to territorial divisions such as roads, hedges, and fields, for isn't mountainous terrain, also in the real world, a domain from which roads and grain fields must withdraw? Combining these two frameworks, we could describe the movement from the premodern to the modern landscape paradigm as a movement from hard mountains, which block land surveying as well as the gaze, to a soft plain that can be measured, plowed, and overlooked.

Indeed, what I have found to be a fertile model is not to demarcate one explanatory "cause," but to coordinate several methodological viewpoints. If Gombrichian perception-cum-schemata thinking seems insufficient today, the method of overcoming it surely isn't to deny—as art historians of the semiotic and deconstructivist bents have tended to do—that artists from the late Middle Ages actually became increasingly oriented toward optical perception.[3] Rather, the challenge consists in demonstrating that optical perception, a perfectly legitimate category, is not merely a matter of innocent looking, but is imbued with ideology; that the emergence of the realist gaze is determined by a cultural transformation reevaluating the very phenomena exposed by that gaze.

More generally, I make use of a macrohistorical, evolutionary model that regards culture as an organic whole built up of interdependent parts. According to the anthropological and sociological theories that have contributed to this model, societies tend to traverse certain universal stages that set limits to how heterogenous a culture can be and, furthermore, presuppose structural correspondences among different cultural domains, if society as a whole is going to function.[4] In the analysis of such domains and their interaction, it has proven useful to include two concepts taken from nonevolutionary theories, Pierre Bourdieu's *field* and Thomas Kuhn's already-mentioned *paradigm*.[5] The *field* signifies a terrain of forces—a space of possibilities—determining what can be realized at a certain historical point in a certain social context. If the field points toward the cultural totality, including its turbulence of contradictory forces, the paradigm hints at its resulting surface: a set of determinable rules such as those governing pictorial space at a certain time.

If we now turn back to our landscape imagery in the light of this methodology, an extended argument could be made, first, that the mountains and rocks of the premodern pictorial paradigm in no way constitute an empty, generalized formula for "landscape," but rather impart resonance to particular culturally significant notions of wilderness.[6] This wilderness can be understood as simultaneously incarnating two existential directions: one horizontal, going outward toward the untouched desert, and the other vertical, going downward toward the chaotic underworld. As the Jungian historian of religions Erich Neumann among others has shown, moving toward the uncultivated desert, especially its mountains, rocks, and clefts, was in fact the same as nearing the primordial chaos that was taken to be concentrated in the womb of the earth mother.[7] The path that Gilgamesh, the Babylonian hero, has to cross to reach paradise beyond the underworld is thus a mountain at the periphery of the world. And in Greek culture, the underworld god Dionysos was celebrated in savage rituals that similarly took place in his home domain, the untouched mountains.[8] What the mountains and rocks of the premodern pictorial paradigm seem to manifest is thus in the broadest sense *terra,* the virgin earth, rather than territory, the controlled and technologically modified earth.

Whether this *terra* points toward the peripheral wilderness or toward the rocky depths of the earth, it is circumscribed with considerable ambiguity. On the one hand, *terra* is destructive chaos, the bowels of Hell that swallow up our lives and grind our bodies to a formless mass. However, just as it consumes it also gives birth, which in other words leads us from an oral-anal to a vaginal function. Here *terra* becomes the creative and nurturing womb of Mother Earth, a regenerative zone that operates as an ideal of origin in the face of civilization's decadence.[9]

The paradise myth, I earlier intimated whose structure might become central for my argument, does so first in relation to this regenerative aspect of *terra.* For the all-nurturing virgin terra is nothing less than a synonym of paradise, the dream landscape where one doesn't have to work. As it happens, the same cultures that introduce the rocky grounds as universal foundations for their landscape depictions—Mesopotamia, Greco-Roman culture, and the European and Byzantine Middle Ages—are also the cultures that mythologize that dream landscape, whether this be the Greco-Roman myth of the Golden Age degenerating into the Silver, Bronze, and Iron Ages or the Judeo-Christian myth of man and woman being expelled from the Garden of Eden.[10] In other words, these cultures formalize a cleft between a primordial epoch in which earth is pervaded with nurturing divinity and a later and more degenerate epoch in which nature has lost its original fertility, so that humans are compelled to work in order to be fed. Whereas the primordial epoch was nurtured by free-growing fruit and, to a certain degree, pastoralism, the degenerate stage is marked by the introduction of grain agriculture.

In fact, the structure of the myth of paradise touches such deeply rooted forces in the premodern Western fields that it is possible to detect a general structural similarity between the narrative course of this myth and the passage of the landscape paradigm in painting from premodern to modern times. To put it briefly, the landscape paradigm experiences a sort of fall from paradise as one moves past 1420. More precisely, both the paradise narrative and the pictorial paradigm encompass a movement (1) from closedness to openness, (2) from timelessness to time, and (3) from virgin earth to tilled ground. Since our interest here lies primarily in the third point, I will only briefly discuss points one and two.[11]

Regarding the movement from closedness to openness: Whereas paradisiac life, because of its lush fruit, is a life of local self-sufficiency, a fact articulated by the wall surrounding the garden of Eden, postparadisiac life is one of dispersion and commerce between distant communities.[12] This movement has its structural similarity in the pictorial paradigm shifting from the more or less shallow spaces of antiquity and the Middle Ages to the infinite perspective space of the paradigm after 1420.

As for the second point, the movement from timelessness to time, paradise denotes an ever-fertile springtime, whereas postparadisiac life is marked by changing and unfriendly seasons. Similarly, the pictorial paradigm of antiquity and the Middle Ages is, as I have already remarked, without traces of time, having a golden or otherwise unchanging sky, whereas the paradigm after 1420 is manifested by cast light; changing, atmospheric skies; and the existence of seasons as well as day-and-night cycles.

Turning now to point three, the movement from virgin earth to tilled ground, let's first consider the situation before 1420. We have already observed how obstinately the pictorial rocks refused to be marked by any traces of the plough, that is, by the technology that arose after the fall from paradise, and, moreover, how they connoted *terra,* that is the paradisiac virgin earth existing as well in the untouched wilderness as in the womb of Mother Earth.

That these rocks could be taken as a terrain for a paradigm that could with any sense be labeled paradisiac becomes especially clear if we examine what sort of human activities are in fact allowed in this paradigm and why. Taking an overview of the surviving images produced from classical antiquity to Carolingian times, it is quite astonishing how few contain renderings of rural labor and technology, especially labor connected to grain culture.[13] In a society in which perhaps 95 percent of the population performed this sort of labor and on the average more than half of the daily food for everybody—high and low—originated from cereals, why is this so?[14] The answer seems to be that rural work, especially work with cereals, was considered degrading by the ruling classes, which longed for, and actively sought to recreate for themselves, a paradisiac, that is, prework, way of life.

In Plato's *Laws,* for instance, we find the telling passage that says that although God no longer governs the world, "We should do our utmost to [. . .] reproduce the life of the age of Cronus," that is, in the Golden Age in which work was unnecessary.[15] In order, however, for the social class of Plato to enjoy sufficient paradisiac leisure so as to perform worthy deeds such as rule, fight, and philosophize, it was necessary that someone else take care of the unpleasant physical work. This was, of course, the workers, not least the rural workers, who in Plato's *Republic,* as in most of Greco-Roman reality, were relegated to a slave class.[16]

To suppress this unpleasant post–Golden Age reality, a longing was directed backward in time toward those occupations that were thought to have existed prior to the discovery of agriculture and which could now, somehow, mirror the leisurely existence of the urban ruling classes. Thus, at the same time classical pictorial space excludes agricultural work, it teems with more primitive and paradisiac occupations, such as herding, fruit gathering, hunting, and fishing. In the so-called sacral-idyllic genre of Roman wall painting, for instance, we meet many fishermen and shepherds but, emphatically, no field workers; and later on, in a *longue durée* deriving from Sumerian times, Christ as the good shepherd is a common figure.

Combining this pictorial obsession with preagricultural activities with the rocky grounds sustaining them, one could remark that at least herding, hunting, and fruit gathering are, in fact, activities that are normal in mountainous areas. Thus, in a more than metaphorical sense, the gaze is led outward in space, to the mountains beyond the cultivated plains.[17] Could one speculate, furthermore, that this rupture of the Egyptian pictorial middle ground toward the more remote, though still not infinitely far, rocky depths presupposes an evolutionary stage in which culture for the first time has become primitivistic—with the result that the very ability to visualize pictorial space through semidistant land formations becomes dependent on a longing (still foreign to less-developed Egypt) toward an unspoiled precivilized stage with a corresponding virgin *terra*?[18]

In any case, although I can merely hint at it here, the opening of the pictorial gaze toward semidistant virgin terrains also implies the emergence of an image genre dedicated to all the features that this gaze overlooks, namely, the *topographical map.* With maps, diagrammatical representations of a gaze looking not outward but rather downward, we enter an ideological space regulated by parameters such as utility, work, control, and colonization—in short, the characteristics banned from the premodern pictorial paradigm.[19] To create a fully perspectival vision, one might speculate that a fusion of the outward-looking and the downward-looking gazes is required, a fusion rehabilitating the ideology of the territory in high art.

Indeed, if we now return to the pictorial paradigm after 1420, we see that the landscape literally experiences a fall. Instead of the mountain slopes with their primitive occupations, we are led down and forward toward civilization's flat plains, on which agricultural fields, hedges, roads, and bridges—those measured divisions and lines through which, furthermore, the new perspective space seems to materialize—become permissible.[20] As Ovid aptly puts it concerning the last of the four world ages, the Iron Age, in his *Metamorphoses:* "And the ground, which had hitherto been a common possession like the sunlight and the air, the careful surveyor now marked out with long drawn boundary-line."[21] Moreover, as an indication that the field divisions are not solely a question of a new and more detailed realism, the paradigm after 1420 is accompanied, although with a certain initial hesitancy,[22] by all the images of rural labor that tended to be absent previously. Think of Bruegel, of Millet and van Gogh, and—perhaps especially—of fascist and communist official art.

If the premodern paradigm with its virgin earth was structurally equivalent to a skepticism toward work on the part of the ruling classes, one could also anticipate that the visualization of the cultivated countryside would similarly be accompanied by a positive attitude toward work among those same classes. This, indeed, seems to be the case. The societal preconditions for such an attitude seem to be constituted, first, by the emergence of a feudal field in the early Middle Ages, and second, by the gradual change of this field into bourgeois capitalism in the centuries following the year 1000. Though feudalism still signifies an aristocratic and hierarchic society, at least the lower classes are metamorphosed from slaves into laborers with a certain formal freedom.[23] As Jacques Le Goff has shown, the potential dignity this lends their work is strengthened by the status of penalty that physical work receives in the medieval cloisters.[24] Correspondingly, from late antiquity onward, although the pictorial paradigm as such is unchanged, images of rural labor scenes become steadily more numerous, accelerating remarkably in the late Middle Ages.[25]

An appropriate forum for such images is the late and postantique genre of *The Labors of the Months*.[26] As is seen from a late example, the one found in the Limbourg brothers' *Les Très Riches Heures du Duc de Berry* from c. 1410–1416—that is, immediately before the breakthrough of the modern pictorial paradigm—such images provide a sort of experimental laboratory for this paradigm.[27] What is called forth here by a certain theme—the monthly labors—soon comes to conquer the landscape grounds in their totality, no matter what iconographic theme is portrayed. Thus, besides the numerous weather situations—snow, sunshine, autumn leaves—one finds in this manuscript some of the first renderings of a terrain permeated with roads, ploughed fields, and fences.

Another example of a pictorial genre exposing prematurely the post-1420 paradigm is *The Good Government in the Country*, Ambrogio Lorenzetti's panorama of Siena's countryside, dating from 1337 to 1340.[28] At stake in this fresco of a city hall's meeting room is the legitimization of Siena's republican government, which, among other things, secures a stable everyday life in the rural uplands of the city. With a bird's-eye perspective that points into the heart of modernity, Lorenzetti has fused the diverse labors of the months—ploughing, sowing, harvesting, etc.—into one industrious season.

What these technologically worked terrains and their heirs since 1420—all the way up to today's supermarket food wrappings—signify, then, is the gradual emergence of a work ethic. The traces of work in the countryside are no longer merely signs of a curse to be hidden away behind a paradisiac facade; on the contrary, they come to acquire moral and even aesthetic value.[29] The work ethic at stake is, in its more mature phase, centered in an urban middle class, creating its worth through a self-made diligence—something that becomes possible in the new market economy emerging in the late Middle Ages.[30] Although this ethic functions most smoothly in capitalism, it can still be regarded as a continuation of what developed behind the cloister's walls. For, as the sociologist Max Weber has shown, work was here designated a calling from God or, as the Germans say, a *Beruf,* and this term extends further, signifying work as well in its more secularized stage. Indeed, according to Weber, the specific capitalist work ethic is not counteracted but rather promoted by religion or, to be more precise, promoted by the different kinds of Protestantism and their forerunners in *devotio moderna,* the northern late medieval movement that tipped Christianity's power center from the Church to the lay congregation.[31]

A convincing indication of this alliance between a capitalist work ethic and Protestantism is seen finally, in the fact that the post-1420 paradigm, with its cultivated landscape—the landscape genre as such—also seems to be especially at home in the Protestant north: The Netherlands, Germany, Scandinavia, northern France, England, the United States. It is here, in the seventeenth to the nineteenth centuries, that portrayals of the countryside—cultivated or untouched—become a genre in their own right.[32]

In the Italy of the Renaissance and the Counter Reformation, on the other hand, an ideal human figure is always at the center of the pictorial vision, with a neo-antique, pastoral landscape serving as discreet background or, at least, as a setting in which the human being is self-evidently enshrined (Figure 3.3). If paradise is less easily abandoned here, a tempting sociological framework might be what Emmanuel Le Roy Ladurie has coined the "immense multi-secural respiration of the social structure," a term referring to the re-feudalization of society which between the late fifteenth and the eighteenth centuries encompassed large parts of Europe, especially its eastern and southern parts.[33] Seen in this perspective, the Renaissance, understood in its literal sense as a movement re-evoking ancient culture, does not emerge as the source of modernity, but rather as a later time pocket, a backlash against the broader field of modernity.[34]

One of the things we could learn from this overview is thus that landscape imagery could be taken as a kind of seismic recorder of the forces pertaining to the social fields. In order to grasp them, one has to move outside the narrow periphery of the iconographic subject matter and into a broader sphere, the iconological, which rather demands a

FIGURE 3.3 *Sleeping Venus* by Giorgione (finished by Titian), c. 1510. Canvas. Courtesy of Gemäldegalerie, Dresden.

structuralist and macrohistorical approach. It is thus through the homologies between different cultural domains that the seismic forces are transmitted.

NOTES

1. On the etymological history of the term, see, for instance, Renate Fechner, *Natur als Landschaft. Zur Entstehung der ästhetischen Landschaft* (Frankfurt, Bern and New York, 1986), 21–23.

2. Ernst H. Gombrich, *Art and Illusion. A Study in the Psychology of Pictorial Representation* (London, 1960).

3. Prime examples of this scepticism toward the perceptual in art are Norman Bryson, *Vision and Painting. The Logic of the Gaze* (New Haven, Conn., 1983), esp. 14–15, 53; and Norman Bryson and Mieke Bal, "Semiotics and Art History," *Art Bulletin 73* (June 1991): 189–190.

4. See especially Talcott Parsons, *The Evolution of Societies,* ed. Jackson Toby (Englewood Cliffs, N.J., 1977) (this book is a synthesis of Parsons's *Societies: Evolutionary and Comparative Perspectives* [Englewood Cliffs, N.J., 1966]

and *The System of Modern Societies* [Englewood Cliffs, N.J., 1971]); Gerhard Lenski, *Human Societies. A Macrolevel Introduction to Sociology* (New York, 1970); and Jürgen Habermas, *Zur Rekonstruktion des historischen Materialismus* (Frankfurt, 1976), esp. 9–48, 129–259. For an overview of the history of evolutionary sociology, see Stephen K. Sanderson, *Social Evolutionism. A Critical History* (Oxford and Cambridge, Mass., 1990).

5. See Pierre Bourdieu, *The Field of Cultural Production. Essays on Art and Literature,* ed. Randal Johnson (New York, 1993), esp. 29–73; and Thomas Kuhn, *The Structure of Scientific Revolutions* (Chicago, 1962), 10 ff.

6. The domination of rocks and mountains in premodern landscape imagery has frequently been observed, but as far as I know, it has never been subject to a systematic interpretation. See, for example, Johannes Jahn, *Antike Tradition in der Landschaftsdarstellung bis zum 15.Jahrhundert* (East Berlin, 1975), 6: "Die Felslandschaft wurde zur Landschaft an sich" (The rock landscape became landscape as such); and Uta Feldges, *Landschaft als topographisches Porträt. Der Wiederbeginn der europäischen Landschaftsmalerei in Siena* (Bern, 1980), 61, 63, who observes "eine tausendjährige Tradition von Felslandschaften" (a thousand-year-old tradition of rock landscapes), combining them with "[eine] klare. Trennung von Felsboden und Einzelpflanzen" ([a] clear separation of rocky ground and single plants). Kenneth Clark, in his still classic *Landscape into Art* (London, 1949), 17–19, first interprets the mountains as "the World beyond the Garden" but then, in considering their survival in the Quattrocento, climbs down: "They survive simply because they were a convenient symbol [. . .]."

7. Erich Neumann, *The Great Mother. An Analysis of the Archetype,* trans. R. Manheim (1955; Princeton, N.J., 1963), esp. 43–46. For this connection among the Jews, see Nicholas Tromp, *Primitive Conceptions of Death and the Netherworld in the Old Testament* (Rome, 1969), 129–144; for Mesopotamia, see Jeremy Black and Anthony Green, *Gods, Demons and Symbols of Ancient Mesopotamia. An Illustrated Dictionary* (Austin, Tex., 1992), 114, 180.

8. For Gilgamesh, see N. K. Sandars, ed., *The Epic of Gilgamesh* (Harmondsworth, 1983), 37, 97–99; for Dionysos, see Karoly Kérenyi, *Dionysos. Archetypical Image of Indestructible Life* (London, 1976), 33–34, 43–45, 210–224.

9. For the different aspects of the earth symbolism, see, in addition to Neumann, *The Great Mother,* James Hillman, *The Dream and the Underworld* (New York, 1979); Mircea Eliade, *Myths, Dreams and Mysteries. The Encounter between Contemporary Faiths and Archaic Realities,* trans. P. Mairet (1957; New York, 1960), esp. 155–170; and Anne Baring and Jules Cashford, *The Myth of the Goddess. Evolution of an Image* (Harmondsworth, England, 1991).

10. For the Greco-Roman myth, see Arthur Ochden Lovejoy and George Boas, *A Documentary History of Primitivism and Related Ideas in Antiquity* (New York, 1935). For the Mesopotamian forerunner, see *Atrahasis* 1 and 7 in Stephanie Dalley, ed. and trans., *Myths from Mesopotamia* (Oxford and New York, 1989), 9, 18. Characteristically, in the more archaic *Athrahasis* myth, work was

originally a phenomenon pertaining to the Gods, who, however, became tired of it and therefore created humans. The idea of an eternally leisurely Beyond belongs to the more advanced evolutionary stage.

11. For a rich documentary evidence of the different *topoi* in Greco-Roman culture, see Lovejoy and Boas, *Documentary History of Primitivism.*

12. *Pairidaeza,* the Persian etymological root of the term *paradise,* suggests in itself a closed entity: As *pairi* is associated with Greek *peri* ('surrounding'), and *daeza* ('clay or dung') is a sticky mass, the word implicates a garden surrounded by a clay wall; see William Alexander McClung, *The Architecture of Paradise* (Berkeley, Los Angeles and London, 1983), 3. For the dispersion of postparadisiac life, see, for instance, Ovid's account of the Iron Age in *Metamorphoses* 1.

13. Several authors have remarked this absence in Greek culture; see Antonio Saltini, *Storia delle scienze agrarie. Venticinque secoli di pensiero agronomico* (Bologna, 1979), 14; Pamela Berger, *The Goddess Obscured. Transformation of the Grain Protectress from Goddess to Saint* (Boston, 1985), 2; and Robin Osborne, *Classical Landscape with Figures. The Ancient Greek City and Its Countryside* (London, 1987), 18–20. Marie-Claire Amouretti, in her *Le pain et l'huile dans la Grèce antique* (Paris, 1986), 293–295, is thus able to find only a little over a dozen surviving examples of Greek plough imagery in the whole time span from the second millenium to the third century B.C.

14. In Peter Garnsey's "Mountain Economies in Southern Europe. Thoughts on the Early History, Continuity and Individuality of Mediterranean Upland Pastoralism," *Pastoral Economies in Classical Antiquity,* ed. C. R. Whittaker (Cambridge, 1988), 198, it is assumed that 65 to 75 percent of the daily average food intake in classical antiquity originated from cereals.

15. Edith Hamilton and Huntington Cairns, eds. *The Collected Dialogues of Plato,* Bollingen Series LXXI (Princeton, 1961), 1305 (*Laws* trans. A. E. Taylor), 731C–E.

16. For the low status of Greco-Roman rural work and its philosophical legitimization, see W. E. Heitland, *Agricola. A Study of Agriculture and Rustic Life in the Greco-Roman World from the Point of View of Work* (Westport, Conn., 1921), esp. 74–77, 92; and Jean-Pierre Vernant, "Arbeit und Natur in der griechischen Antike," *Seminar: Die Entstehung von Klassengesellschaften* (Frankfurt, 1973), 246–270.

17. In his macrohistorically tuned *The Mediterranean and the Mediterranean World in the Age of Philip II* (London, 1972), 1:30 ff., Fernand Braudel stresses, from a sociological point of view, how the mountain cultures can be seen as an origin for Mediterranean civilization.

18. For Egypt's pertaining to an earlier evolutionary stage than the Mesopotamian city states, see Parsons, *Evolution of Societies,* 68. As an indication that this stage is in fact less marked by a longing toward an unspoiled past, Egypt's paradise, Earu's fields, still had to be worked in order to be fertile; see Wolfgang Helck et al., ed., *Lexikon der Ägyptologie,* vol. 1 (Wiesbaden, Germany, 1975), col. 1156: "Earu-Gefilde"; and Erwin Panofsky, *Tomb Sculpture.*

Four Lectures on Its Changing Aspects from Ancient Egypt to Bernini (New York, 1964), 14–15.

19. For a discussion of Roman agricultural maps, see O. A. W. Dilke, *The Roman Land Surveyors. An Introduction to the Agrimensores* (Newton Abbot, UK, 1971).

20. Although arriving at it by a different, more empirical road, it is this kind of repertoire Martin Warnke hints at with his concept of "political landscape." See his *Political Landscape. The Art History of Nature,* trans. D. McLintock (1992; London, 1994), esp. 9–20 ("The Occupations of the Plain").

21. Ovid, *Metamorphoses,* trans. F. J. Miller (Loeb Classical Library: London, 1916), 1:11, 13.

22. The apparent subduing of agricultural imagery in Dutch seventeenth-century landscape painting could be taken as a good example of this hesitancy—although it should be stressed that in comparison with previous developments, Dutch art of this century in fact signifies not a subduing but an evoking of imagery connoting work.

23. For the postantique upgrading of the status of work, see Heitland, *Agricola,* 207–212; for an intriguing neo-Hegelian explanation of this, see Alexandre Kojève, *Introduction to the Reading of Hegel. Lectures on the Phenomenology of Spirit,* comp. R. Queneau, ed. A. Bloom, trans. J. H. Nichols (1947; New York, 1980).

24. Jacques Le Goff, *Time, Work and Culture in the Middle Ages,* trans. A. Goldhammer (1977; Chicago and London, 1980), 71–86.

25. This acceleration is mostly remarked in Marxist analyses. See Siegfried Epperlein, "Bäuerliche Arbeitsdarstellungen auf mittelalterlichen Bildzeugnissen. Zur geschichtlichen Motivation von Miniaturen und Graphiken vom 9.bis zum 15.Jahrhundert," *Zeitschrift für Wirtschaftsgeschichte,* (1976): 181–208, and *Der Bauer und seine Befreiung. Kunst vom 15.Jahrhundert bis zur Gegenwart* (Dresden, 1975), esp. 26. See further Emmanuel Le Roy Ladurie, ed., *Paysages, paysans. L'art et la terre en Europe du Moyen Age au XXe siècle* (exhibition catalogue) (Paris, 1994).

26. James Carson Webster, *The Labors of the Months in Late Antique and Medieval Art. To the End of the Twelfth Century* (Princeton, N.J., 1938).

27. *The Trés Riches Heures of Jean, Duke of Berry, Musée Condé, Chantilly,* introduction and legends by Jean Longnon and Raymond Cazelles with preface by Millard Meiss (Secaucus, N.J., 1969).

28. See for instance Feldges, *Landschaft als topographisches Porträt,* Fechner, *Natur als Landschaft,* 209–228, and Diana Norman, ed., *Siena, Florence and Padua. Art, Society and Religion 1280–1400* (New Haven, Conn., and London, 1995), 2:145–167.

29. Although somewhat downplaying the moral aspect, Joachim Ritter, in his 1963 essay "Landschaft. Zur Funktion des ästhetischen in der modernen Gesellschaft" (*Subjektivität. Sechs Aufsätze* [Frankfurt, 1989], 141–190) includes landscapes that have been technologically modified as examples, for such an aesthetic feeling for nature.

30. For the gradual emergence of a capitalistic economy from the twelfth century onwards, see, for instance, Fernand Braudel, *Civilization and Capitalism 15th–18th Century,* vol. 2, *The Wheels of Commerce,* trans. S. Reynolds (1979; Berkeley and Los Angeles, 1992), 232–233; and George Ovitt, *The Restoration of Perfection. Labor and Technology in Medieval Culture* (New Brunswick and London, 1987), 139.

31. Max Weber, *The Protestant Ethic and the Spirit of Capitalism,* trans. T. Parsons (1920; New York, 1950).

32. For an analysis of the emergence of the autonomous landscape genre and its relation to a northern European sensibility, see Christopher S. Wood, *Albrecht Altdorfer and the Origins of Landscape* (London, 1993).

33. Le Roy Ladurie, *Paysages, paysans* 18: "Cette immense respiration multiséculaire d'une structure sociale." Further, Braudel, *The Wheels of Commerce* 265–72.

34. For the idea of the Renaissance as a backlash, see Oswald Spengler's splendid observations in his *Der Untergang des Abendlandes. Umrisse einer Morphologie der Weltgeschichte* (1923; Munich, 1972), esp. 300–308.

4

The Employment of the Word

Writing, Topography, and Colonial Landscapes

TADEUSZ RACHWAL

Take, for example, the plants of the genus *Convolvulus,* trailing plants of the morning-glory family with funnel-shaped flowers and triangular leaves. In 1576 the French Botanist Charles de Lécluse . . . designated one species as *Convolvulus folio Altheae.* In 1623 the Swiss Botanist Gaspard Bauhin . . . called this same species *Convolvulus argenteus Altheae folio,* which in 1738 Linnaeus amplified to *Convolvulus foliis ovatis divisis basi truncati: laciniis intermediis duplo longioribus,* which by 1753 he had elaborated further into *Convolvulus foliis palmatis cordatis sericeis: lobis repandis, pedunculis bifloris.*

—Daniel J. Boorstin, *The Discoverers*

The most general definition of *technology* to be found in *Webster's Collegiate Dictionary* is "the totality of the means employed to provide objects necessary for human sustenance and comfort." *Encyclopaedia Britannica,* interestingly, defines it as "the systematic study of techniques for doing things" (combination of the Greek *techne* and *logos,* 'art' and 'word'). In the encyclopedia's long entry for *technology,* which provides a history ranging from the bow and arrow, food production, and soap making to space probes, the art of the word goes unmentioned. Given that human speech might still be regarded as a part of human nature, writing, quite evidently, is something that can "do things" and thus should also be viewed as a technology of sorts. It can also "provide objects" as well as deprive us of them. In this essay, I focus on the ways the "employment" of language and writing produced something that might be called a reduction of landscape in colonial and scientific discourses of the eighteenth century.

Naming newly discovered territories and mapping them are obvious examples of such an employment, and, in a certain sense, geographical naming is a production of places as objects, as areas available both physically (they must have been achieved, or discovered, before they were mapped) and ideologically (as the areas of jurisdiction and administration they become the moment a name is given to them, in the same way a baptized person becomes a member of a church). There are, it seems, no maps without names, and maps' primary function is not only to show *where* the named places are. First of all they tell us *that* they are, and, secondly, they tell us *whose* they are. Hence, there is a paradox involved in the very idea of place, which belongs to both the sphere of seeing and the sphere of writing. As Georges van den Abbeele notices, "A place can only 'take place' within a text, that is, only if it can be marked and re-marked from the area in which it is inscribed. Only in this sense can we speak of a topo*graphy,* for insofar as the very perception or cognition of a landscape requires an effect of demarcation, the latter can only be constituted as a space of writing."[1]

Though mapmaking started from "the self-evident boundaries of land and sea," it later became necessary to inscribe the artificial boundaries of latitude and longitude upon them as a grid enabling finding and reading places as systematically related to each other. Thus, perhaps paradoxically, it is "maps that generate movements, and not the reverse."[2] Paul Foss notes that "map-making is image making." It is "pure performance, and in that sense it is always creative of itself. Which is why maps are projections in space of strategies and maneuvers to come. The map of Cook, therefore, does not represent so much the discovery of a territory called New South Wales, but rather projects the grid within which all sorts of territorial movements had already begun to circulate—and which, precisely because of the map, could now circulate on a greater scale."[3] A map is a promise of a landscape which it cannot represent and an invitation to seeing it, to traveling to places already parceled and allotted. In colonial discourse, this promised landscape is usually rendered as new, exotic, and different, though still familiar exactly as landscape, as a view to be seen whose novelty and virginity, in turn, stir an excitement to possess, though from an epistemologically secure, because mapped, position. Discoverers are mapmakers, regardless of whether they discover and name new worlds; new objects; or, like botanists, new species. Certain mental landscapes are also quite evidently areas of discovery.

In his nostalgic view of the innocence of Indians due to their ignorance of writing (an idea that Gordon Brotherston perceives as a kind of strategic omission, if not a lie), Claude Lévi-Strauss, in *Tristes Tropiques,* presents the savage mind as a kind of otherness inaccessible to the linear technology of the European perception of things and places.[4] By doing so, he posits "savagery" as the lost originating innocence whose distortion by the supplement of writing makes it irrevocably alien and inaccessible to any kind of speculation. Eurocentrism translates any rendition of that innocence as an always already mirrored image of the linearity of our thinking. Like Foss's mapmaking, Western epistemology is image making, a production of things and their representations in the image of Europe. Like the truth of the metaphysics of presence, on the other hand, the truth of the savage mind lies beyond writing, beyond the technology of mapping whose employment of the word, as it were, deprives Indians of the ability of "doing things with words," of having places—a strategy present already in the early colonialists' depriving the conquered peoples of private property, ethics, and even language. Lévi-Strauss's radical othering of the savage mind equally radically shifts true experience and immediate knowledge of the world beyond the reach of philosophy and science, nostalgically projecting an ideal society in which, as in Baudrillard, we are all Indians,[5] though we have forgotten about it. Perhaps somehow paradoxically, the image of truth or, better, of its structure seen as the beyond of writing, remains, in Lévi-Strauss, as metaphysical as in, say, Plato.

Indians are structurally European within. Indians in fact live structurally, beyond the secondary actualization of structures in writing. A similar conviction was a mechanism of much of the colonial linguistic technology of conversion, in which the certainty that there was a universal language of truth underlying all expression made the colonizers concentrate on the sphere of performance, actualization, whose mastery was seen as a trait of humanity and innate Christianity. The difference of performance, which Lévi-Strauss conceals as the nonexistent Indian writing, was a symptom of Indians' inferiority and an invitation to guidance. The same seems to hold for the practice of mapmaking, an activity not quite unknown to Indians.

In the eighteenth century, Indians' cartographic skill seems to have been an object of admiration on the part of some white men as an innate, natural gift to depict the landscape. Governor James Glen of South Carolina, for example, saw Indians' cartographic skill as almost equal to that of the Europeans: "I have not rested satisfied with a verbal description of the country from the Indians but have often made them trace the rivers on the floor with chalk, and also on paper, and it is surprising how near they approach to our best maps."[6]

When it comes to some decision making, however, such as establishing the boundary between the colonies and the Indian territories, a matter on which "talks" were held in the 1760s and 1770s, the mistrust as to that innate ability clearly surfaces. In a report of 1766 concerning the demarcation of the boundary between the Cherokee and South Carolina, we read, "The Cherokee propose running the line from where it terminated a straight course, to Colonel Chiswell's Mines, which I believe will be north, as nigh as I could make it: . . . It would be very necessary, that a surveyor should first sight the line, from Reedy River a north course, in order to know where it will terminate in Virginia, and whether or not, it will take away any of the settlements."[7] The Cherokee proposal is not quite clear to the author of the report ("as nigh as I could make it"), and the suggested line needs to be surveyed; it demands an expertise not so much as regards the idea of the boundary, whose status is seen as universal, but as regards the cartographic performance of Indians, in which such places as Virginia and Colonel Chiswell's mines must be considered natural topographic elements of the environment. Virginian landscape also already contains settlements whose taking away from the properly British territory might deprive the map of a few names whose very presence there determines the places' belonging.

William Bartram, a British botanist and a lover of nature living in America, traveled to the Indian territories in the 1770s. What he produced were not exactly maps but descriptions of landscapes of the Indian land from the seemingly impartial perspective of an observer and a scientist. His reflections upon seeing the landscape there are first of all economic rather than aesthetic. He writes, for example, "This new ceded country promises plenty and felicity . . . The hills suit extremely well for vineyards & olives as nature points out by the abundant produce of fruitful grape vine, native mulberry trees of an excellent quality for silk. Any of this land would produce indigo & no country is more proper for the culture of almost all kinds of fruit."[8] Bartram describes here the agricultural potential of the landscape, rather sparing us the description of what it looks like. The land having been made accessible, it now awaits improvement and correction in the manner it needs familiar names in the eyes of a cartographer. Landscape, much like the eighteenth-century landscape garden, is posited as a space for what should be rather than what is. The country, though uncultivated, is prospectively proper for cultivation, this property promising felicity when agricultural technology replaces its imperfect construction. Bartram's survey in fact "names" what is not there, familiarizes the landscape as the area that only wants some labor to become a land of plenty, which it as yet is not.

If cartographers name and order the world, botanists, like Bartram, attempt to map nature. According to Daniel J. Boorstin, two "great systematizers—[John] Ray and Linnaeus—would accomplish for all plants and animals what Mercator and his fellows did for the planet's whole surface." In botany, what is needed instead of places are species as stable elements of categorization whose exemplary adequacy of naming I have quoted in the epigram in the beginning of this essay. If Ray was, say, the founder of species as a way of seeing plants and animals (the word *species* also derives from *specere,* 'to look at,' 'to see') in groups rather than as a landscape, for example, Linnaeus voyeuristically looks at the species' intimate parts as the underlying foundation of his system in which the number of stemens in flowers indicates the number of husbands (e.g., *Diandria*—two husbands in the same marriage). Where there seem to be no flowers, the plants are "cryptogamic" and have only "female" organs. As Boorstin notes, "even an accomplished botanist like the Reverend Samuel Goodenough . . . who had a plant, goodwinia, named after him, could not conceal his embarrassment at the 'gross prurience of Linnaeus' mind . . . A literal translation of the first principles of Linnaean botany is enough to shock female modesty.' It is possible that many virtuous students might not be able to make out the similitude of Clitoria."[9]

In Bartram's *Travels of William Bartram,* the employment of the word becomes exactly a Linnaean kind of enterprise in which the landscape he sees is transformed into a display of species. Bartram's father, John, was Linnaeus's correspondent in Philadelphia. He gives us glimpses of the sublimity of the landscape in order to desublimate it almost at once and provide us with a list of names of the species, probably recognized by a close look at their sex organs. For example, having reached the top of the Occonne mountain, Bartram rests for a while and informs us that:

> I was now in a very elevated situation, from whence I enjoyed a view inexpressibly magnificent and comprehensive. The mountainous wilderness which I had lately traversed, down to the region of Augusta, appearing regularly undulated as the great ocean after the tempest; . . . My imagination thus wholly engaged in the contemplation of this magnificent landscape, infinitely varied, and without bound, I was almost insensible or regardless of the charming objects more within my reach: a new species of Rhododendron foremost in the assembly of mountain beauties; next the flaming Azalea, Kalmia latifiola, incarnate Robinia, snowy manteled Philadelphus inodorus, perfumed Calycanthus, &c.[10]

The "&c." clearly points to the dream of finalizing the naming, so that even the "charming objects" outside the reach of Bartram's classificatory capacity (i.e., of his eventually naming everything within the landscape "without bound" that constitutes the background for the naming of the new species) somehow erotically invite his gaze and expose their, say, secret names along with their parts. The mountain beauties, such as flaming Azalea, are thus familiarized; they simultaneously become members of a family of plants and part of the corpus of European knowledge by way of coming out to the foreground from the sublime, sealike chaos of the landscape.

Bartram, of course, was not alone in his enterprise as a "naturalist naturalizing the bourgeois European's own global presence and authority," as Mary Louise Pratt phrases it. Linnaeus himself, in a letter to his colleague written in 1771, seems to assume the position of an emperor of nature whose ambassadors keep sending reports and trophies to him in Sweden from all over the planet:

> My pupil Sparrman has just sailed for the Cape of Good Hope, and another of my pupils, Thunberg, is to accompany a Dutch embassy to

> Japan; both of them are competent naturalists. The younger Gmelin is still in Persia, and my friend Falck is in Tartary. Mutis is making splendid botanical discoveries in Mexico. Koenig has found a lot of new things in Tranquebar. Professor Friis Rottboll of Copenhagen is publishing the plants found in Surinam by Rolander. The Arabian discoveries of Forskal will soon be sent to press in Copenhagen.

In Copenhagen the nature of the world is translated, or "pressed," into the order of the book. Rottboll's plants from Surinam get published as a part of the global project of mapping the world and ordering it through the authority of print. Natural history supplements navigational mapmaking by mapping out, as Pratt notes, "not [the] thin track of a route taken, nor the lines where land and water meet, but the internal 'contents' of those land and water masses whose spread made up the surface of the planet." Immense landscapes, as in Bartram, get divided and minutely penetrated so as to enable the mapping of "every visible square, or even cubic, inch of the earth's surface."[11] The sublime terror of vastness and infinity is overcome by natural historians through particularization and diminution, a gesture constitutive of beauty, at least in Burke.[12] Natural history in fact aestheticizes the world, showing it as attractive because possessable in the form of small objects that one can send to a Copenhagen press, for example. Without natural history, the world is prodigious and diversified to the point of Bartram's loss of sensitivity in the face of the infinite variety of the landscape. The technology that is supposed to reproduce such an immensity must thus also be immense, and immensity is exactly the word with which Buffon, in 1749, characterized natural history: "Natural history taken in its full extent, is an immense History, embracing all the objects that the Universe presents to us. This prodigious multitude of Quadrupeds, Birds, Fish, Insects, Plants, Minerals, etc., offers a vast spectacle to the curiosity of the human spirit; its totality is so great that it seems, and actually is, inexhaustible in all details."[13]

Bartram, as we have seen, not only classifies the world according to species, but also judges it in terms of the potential growth of plants to the benefit of humankind and the state. An emissary of both natural history and the state, he depicts virgin American landscapes as populated mostly with Latin names of plants, leaving the prodigious multitudes at a distance. One could easily imagine that what grows there, at a closer look, are words. Bartram's America is already ordered and secure, a naturally historicized territory in which everything is in its place and from which the approaches of sublimity have been pushed away. If, according to Michel Foucault, natural history reduces "the whole area of the visible to a system of variables all of whose values can be designated,"[14] then the things reduced, the unclassified and unclassifiable ones, obviously form an unwelcome part of the whole project, regardless of the fact that they also, as in Buffon, promote it.

Bartram's pen reduces the natural landscape to a botanical garden, an ordered, though linguistic, visual space whose technological transformation into actual spaces gave occupation to numerous scholars in eighteenth-century Europe. If publication of a book on natural history provided a discursive frame for the creation of scientific authority, a botanical garden was an ideal space for publicly displaying the discovered order of nature as actual and real. Hence, as Pratt observes, "botanical gardens became large-scale public spectacles, and the job of supervising them a naturalist's dream. (Buffon became keeper of the king's garden in France, while Linnaeus devoted his life to his own.)"[15] Bartram, it seems, wanted to write a book that simultaneously was a botanical garden. He wanted to press the American landscape and plant it in the pages of his *Travels*.

In Bartram, the Linnaean "dogma of sexuality," as Goethe had it,[16] also translates landscape into an already fully possessed domain, into meadows full of flowers that expose their organs to the curious gaze of the British visitor, revealing the concealed truth of their species. When at one point in the book he sees Indian women, it is difficult to tell whether he is not still describing animals and plants—though, surprisingly, he does not endow the women with a Latin name. Walking with his companion, a trader, Bartram notices

> a most enchanting view, a vast expanse of green meadows and strawberry fields, a meandering river gliding through . . . green turfy knolls, embellished with parterres of flowers and fruitful strawberry beds; flocks of turkeys strolling about them; herds of deer prancing in the meads or bounding over the hills; companies of young, innocent Cherokee virgins, some busy gathering the rich fragrant fruit, others having filled their baskets reclined under the shade of floriferous and fragrant native bowers of Magnolia, Philadelphus, perfumed Calycanthurs, sweet Yellow Jassamine and cerulean Glycine frutescences.

Hidden behind a bush, the two explorers now observe the "landscape," seeing it as a spectacle of nature displaying itself, innocently laying bare its "beauties" and thus proving to be naturally seductive. The Cherokee women are described as "disclosing their beauties to the fluttering breeze, and bathing their limbs in the cool, fleeting streams; while other parties, more gay and libertine, were yet collecting strawberries, or wantonly chasing their companions, tantalising them, staining their lips and cheeks with the rich fruit." "Nature prevailing over reason," Bartram informs us, the two impartial observers (perhaps prompted by the affinity of the species) decide to reveal themselves and take a more active part in the scene, which was "perhaps too enticing for hearty young men to long continue idle spectators." The women, "on perceiving themselves to be discovered by us, kept their station, peeping through the bushes . . . when observing our approaches, they . . . decently advanced to meet us, half unveiling their blooming faces, incarnated with the modest maiden blush, and with native innocence and cheerfulness, presented their little baskets, merrily telling us their fruit was ripe and sound."[17] Seeing nature is always a discovery in Bartram, and this peculiar discovery of women as innocent pickers of fruit translates implicit eroticism into an explicit economy of natural exchange in which the discovery itself is an act of taking fruit that nature voluntarily offers.

Perhaps behind natural historians' interest in flowers and sexual organs there lies an equally strong interest in production, in the provision of fruits and the preservation of their likeness, which could be guaranteed only by the fixity of the species, which in turn guaranteed the fixity of production and reproduction. The Linnaean Latin-like nomenclature was a tool that created such a fixity on the level of language, whose changeability and differences in the case of natural languages did not guarantee that biologists were talking about the same things. With the coming of Linnaeus's system, nature could eventually be listed, could emerge from the creation—"Deus creavit, Linnaeus disposuit" as his admirers used to say—and thus become "objects necessary for human sustenance and comfort," which task, as *Webster's* tells us, can be achieved only by means of a technology. Natural historians' employment of the word was actually a new creation of nature, its translation into a new book of the world that, hand in hand with the encyclopedia, was now displaying the world, or at least one of its domains, as eventually mastered.

NOTES

1. Georges van den Abbeele, *Travel as Metaphor. From Montaigne to Rousseau* (Minneapolis: University of Minnesota Press, 1992), xxi.

2. Daniel J. Boorstin, *The Discoverers* (New York: Random House, 1983), 432.

3. Paul Foss, "Theatrum Nondum Cognitorum," Folio of Art and Design (Winter, 1980), 20.

4. See Claude Lévi-Strauss, *Tristes tropiques* (Paris: Plon, 1990). See also Gordon Brotherson, "Towards a Grammatology of America: Lévi-Strauss, Derrida, and the Native New World," in *Europe and Its Others,* ed. Francis Barker, Peter Hulme, Margaret Iversen, Diana Loxley, vol. 2 (Colchester, England: University of Essex, 1985).

5. See Jean Baudrillard, *L'échange Symbolique et la Mort* (Paris: Gallimand 1976), 36: "There will always be animal and Indian reservations to conceal that they are dead, and that we are all Indians." For a more detailed discussion of this idea, see Tadeusz Rachwal, "History. Of Indians," in *'We are all Indians.' Violence, Intolerance, Literature,* ed. Tadeusz Slawek and Wojciech Kalga (Katowice, Poland: Uniwersytet Slaski, 1990).

6. William L. McDowell Jr., ed., *Documents relating to Indian Affairs,* May 21, 1750–August 7, 1754, South Carolina Archives Department, Columbia, S.C., 1958, 536.

7. British Public Record Office, *Colonial Office Records, America and the West Indies, 1601–1807,* vol. 66 (London: British Public Record Office, 1979), 398–399.

8. William Bartram, "Travels in Georgia and Florida, 1773–1774: A Report to Dr. John Fothergill," *Transactions of the American Philosophical Society* 33 (November 1943): 144.

9. Boorstin, *The Discoverers,* 432, 434, 439.

10. Mark van Doren, ed., *The Travels of William Bartram* (New York: Dover Publications, 1928), 273–274.

11. Mary Louise Pratt, *Imperial Eyes. Travel Writing and Transculturation* (London: Routledge and Kegan Paul 1992), 28, 27, 30.

12. Edmund Burke, *A Philosophical Enquiry into the Origin of Our Ideas of the Sublime and Beautiful* (Oxford: Oxford University Press, 1990), 36.

13. Pratt, *Imperial Eyes,* 30.

14. Michel Foucault, *The Order of Things* (New York: Pantheon, 1970), 136.

15. Pratt, *Imperial Eyes,* 29.

16. See Boorstin, *The Discoverers,* 439.

17. van Doren, *Travels of William Bartram,* 289.

5

Remaking a "Natural Menace"

Engineering the Colorado River

DAVID E. NYE

Today's visitor to the Grand Canyon and the lower Colorado River region arrives with a set of cultural predispositions that seem far removed from those of explorers and early settlers.[1] Because the Grand Canyon has become sacrosanct to most Americans and the most popular of the national parks,[2] it is important to remember that John Wesley Powell's early expeditions did not set out with the goal of discovering such a shrine. Exploration was assumed to be the first stage before a region's exploitation and settlement. Present-day New Mexico, Arizona, California, Utah, and Nevada had belonged to Mexico until a decade before the Civil War, and much of the area remained terra incognita to Anglo-Americans. The public and the government perceived the Colorado River valley through the lens of Manifest Destiny, and the intentions of early explorers in this region were much the same as those who had been first on the Hudson, Ohio, Mississippi, Missouri, and Columbia Rivers. These rivers were the natural gateways to new regions. The expectation was that along their banks farmers would settle and towns would rise.

Navigable rivers were the key to regional development. The first federally funded expedition on the Colorado River began in the Gulf of California and explored upstream by steamboat, to see how far one

I want to thank the Center for Man and Nature, Odense University, for a research grant in 1996 for a larger project on the landscapes of the lower Colorado River, of which this essay forms a part.

could go without serious impediment. The federal government, laissez-faire in most areas, was active in clearing river waterways. Making them safe for navigation was an obvious boon to development. A few steamboats had already begun operating on the lower Colorado in 1852. Five years later, Lieutenant Joseph Christmas Ives was instructed by the War Department to explore further. He did so in a boat manufactured for that purpose in Philadelphia. After it was tested on the Delaware River, the boat was disassembled, shipped to the mouth of the Colorado, and put together again. The fifty-foot stern-wheeler was ready in December and spent the next two months beating its way upstream. Often stuck on sandbars, the boat managed to reach Black Canyon, near the present site of Hoover Dam, before suffering damage by striking a sunken rock.[3] The boat was left behind for repairs and eventually sent downstream again,[4] while the expedition proceeded on foot into the lower end of the Grand Canyon, bringing back the first American descriptions and drawings of the site. Ives made his way a considerable distance up the canyon; passed Diamond Creek; and, with Hualupai guides, climbed more than five thousand feet up to the rim of the canyon. The party descended with great difficulty into several side canyons and found its way constantly blocked by vast natural obstacles. From a position somewhere on the rim, Ives recorded:

> The region east of camp has been examined today. The extent and magnitude of the system of cañons in that direction is astounding. The plateau is cut into shreds by these gigantic chasms, and resembles a vast ruin. Belts of country miles in width have been swept away, leaving only isolated mountains standing in the gap. Fissures so profound that the eye cannot penetrate their depths are separated by walls whose thickness one can almost span, and slender spires that seem tottering upon their bases shoot up thousands of feet from the vaults below . . . Our reconnoitering parties have now been out in all directions, and everywhere have been headed off by impassable obstacles. The positions of the main watercourses have been determined with considerable accuracy. The region last explored is, of course, altogether valueless. It can be approached only from the south, and after entering it there is nothing to do but to leave. Ours has been the first, and will doubtless be the last, party of whites to visit this profitless locality. It seems intended by nature that the Colorado river, along the greater portion of its lonely and majestic way, shall be forever unvisited and undisturbed . . . The country could not support a large population.

Whereas Ives judged the Colorado River gorge to be "profitless" and "valueless," he devoted several pages to the land of the Moqui tribe, much of which was cultivated by irrigation, and which he called "the most promising looking for agricultural purposes of any yet seen."[5]

However, white settlements in the region would not be easy, for the Colorado River was difficult to navigate. The accompanying "Hydrographic Report" stated that in the delta region, "memory assists but little in selecting the channel, for it has been known to change from one bank to the other in a single night. The water being turbid and perfectly opaque, it is impossible to determine the depth as in the case of a clear stream." Only an experienced pilot could manage the job. "From the formation and relative positions of the islands and banks, from the eddies, the direction of the currents, from the pieces of driftwood, and other floating substances, the experienced navigator can generally determine the proper course." Farther upstream, the river contained rocks, rapids, and rough gravel shoals but was generally navigable. The farther north one went, the shallower the river generally became, and at Black Canyon, near the present site of Hoover Dam, navigation became so difficult that it was not to be contemplated except with moderately high water. Given these conditions, the expedition party suggested that a 100-foot stern-wheel steamboat made from iron, with a draught when light of not more than twelve inches, be built for service on the Colorado, from the delta to Black Canyon. They estimated that such a boat could make about fifty trips in six years, transporting three thousand tons of freight at a cost of about $30 a ton.[6] As these and other calculations in his report suggest, Ives believed that the Colorado River valley should and could be settled and exploited.

After explorers came surveyors, and indeed some later expeditions performed both functions. Surveyors were the essential precursors of settlement, providing the lines that divided the land into a grid, according to repeatable scientific procedures. For example, a decade after Ives's trip, Clarance King surveyed a band 100 miles wide along the fortieth parallel, between Nevada and Colorado. This area included the route of the new Central Pacific Railroad that formed part of the first transcontinental railway, completed in 1869. King published an impressive report, *Mining Industry*, in 1870, suggesting the rich economic possibilities of this region. He was thus both a surveyor and a promoter, by no means an unusual combination. More than half of his report focused on the Comstock Lode in Virginia City, Nevada, which had produced immense quantities of silver since being discovered in 1859. Mark Twain published a popular account of these mining camps in *Roughing It* the following year.[7]

Even before King's report appeared, Lieutenant George Montague Wheeler, of the Army Corps of Engineers, was sent from San Francisco to Nevada and the Colorado River area in 1869, leading a party of thirty-six people, seventy-nine mules and horses, and eight wagons. His mission was not a detailed survey but a reconnaissance, and "the focus of the entire five-month expedition" was "primarily mining districts, their access to transportation routes, and the disposition of the Nevada Indians." At the same time, the Treasury Department had "a special commissioner of mining statistics, Rossiter Raymond," who was "covering the same mining districts and providing in annual reports the same kinds of information."[8] Henry Adams noted in 1871 that the pursuit of gold and silver in the Great Basin created "a feverish, or even furious, spasm of speculative excitement, during which roads and stage lines traversed the region in every direction, while mining camps, towns, and even cities shot up like mushrooms."[9]

These surveyors, travelers, and prospectors were not immune to the pleasures of landscape. Wheeler appreciated the rugged scenery according to the accepted aesthetic of the time, calling the Colorado River canyon "the most wild, picturesque and pleasing of any that it has ever been my fortune to meet."[10] King had loftier sentiments, cultivated by a wide reading, knowledge of art, and even personal acquaintance with John Ruskin, who once gave him two paintings by Turner.[11] Powell, whose descriptions of the Colorado River still inspire preservationists,[12] nevertheless spent years promoting dam construction and irrigation of the West, although he understood that much of the region, though fertile, could never be supplied from the scant water sources available.[13] As Donald Worster has noted, "Powell was deeply moved by the sublime geology and brilliant colors of the canyonlands, by sweeping desert vistas, by mountain forests and rills." Yet Powell also "favored turning as much of the wilderness into 'productive' agricultural landscapes as possible—and so was virtually every other white man of his age, except John Muir."[14]

Native Americans, in contrast, did not see the region in abstract terms, whether the numerical abstractions of the surveyor or the aesthetic abstractions of the landscape artist. Powell remarked,

> It is curious now to regard the knowledge of our Indians [i.e., the guides]. There is not a trail but what they know; every gulch and every rock seems familiar. I have prided myself on being able to grasp and retain in my mind the topography of a country; but these Indians put me to shame. My knowledge is only general, embracing the more important features of a region that remains as a map engraved on my mind; but theirs is particular. They know every rock and every ledge, every gulch and cañon, and just where to wind among these to find a pass; and their knowledge is unerring. They cannot describe a country to you, but they can tell you all the particulars of a route.

Native Americans did not create systems of abstraction to comprehend the appearance of the land in general patterns. Powell saw the landscape in terms of geology and devoted pages of his government report to classifying the different forms of desert valleys, first in the abstract (diaclinal valleys, which pass through a fold in sedimentary beds; synclinal valleys, which follow synclinal axes; etc.) and then in the particular cases of the Green and Colorado Rivers and their tributaries.[15] Combining this sort of knowledge with his appreciation of landscape, Powell created a "map engraved on my mind." This sensibility, which linked appreciation of the sublime with the new science of geology, could even be made congruent with a rather pious interpretation of nature.[16] A generation earlier, some Americans still saw Niagara Falls as proof against atheism, but most subsumed it within the aesthetic of the sublime.[17]

Geology complicated such formulations but hardly made them impossible. In the case of the Grand Canyon, the geologist Clarence Dutton wrote a detailed study of its rock formations, yet retained a reverential awe toward the site as a whole. As Stephen J. Pyne put it, "For Dutton, the scenic cliff and the geological cliff were the same cliff. Both could be understood by similar acts of critical imagination, and one could be trusted to reinforce the meaning of the other."[18] Dutton was able to put into words an enthusiasm many felt but could not express, in a passionate scientific analysis. Even before his verbal descriptions were published, the sublime vision of the canyon had been embodied in Thomas Moran's large format painting *The Chasm of the Colorado,*

1873–1874, which was immediately bought by Congress for $10,000 and displayed in the Capitol dome.[19] This vision was further expressed in other Moran paintings, many of them purchased, reproduced, and distributed nationally by the Santa Fe Railroad after 1893.[20]

In short, whereas Native Americans knew every nook and cranny of the land as a series of details that they often imbued with spiritual significance, for most Americans a feeling for the sublime and the impulse to exploit the region coexisted. Wheeler admired the scenery but also made special note of good agricultural land; the possibilities for irrigation; and, most of all, existing and potential mines. In 1871 he led a second expedition into the Grand Canyon area, taking along photographer Timothy O'Sullivan, who produced not only some of the most famous early photographs of the canyon, but also images of mines, including smokestacks, shafts, and mounds of tailings. As these latter images suggest, Wheeler included in his various expeditions not only photographers but investors from San Francisco, whose names soon appeared on many land claims in Arizona and Nevada.

If these early explorers did not always separate private and public interests, they had even more difficulty recognizing Native American claims to the land. Wheeler, a recent graduate of West Point who had finished first in his class in philosophy and mathematics and second in engineering, hardly held enlightened views.[21] In a report in 1879 he attacked what he called the "peace-at-any-cost-policy" that the federal government was pursuing with Native Americans and declared, "The ever-restless surging tide of population, almost a law unto itself, already in many cases crowds over the borders of these [Indian] reservations, and the time is not far distant when the question of the surrender of these lands to actual settlers will naturally be answered in the affirmative, on the plea of the greatest good for the greatest number." Such views were widespread, and the appeal to "the greatest good for the greatest number" was the constant utilitarian justification. In Wheeler's view, the exploitation of land and disinheritance of its original inhabitants provided a suitable topic "for the ethnologist and philosopher, but scarcely for the practical man of affairs, intent on wresting from productive nature the largest bounty."[22] For Wheeler, bounty usually meant mining. In 1872 alone, his survey examined forty-eight mining districts in Utah, Arizona, and Nevada.[23] For others, the bounty of the West might mean lumbering, irrigated farming, or ranching, but in all cases it meant individuals taking possession of the land. Creating a national park in the major river valley penetrating the region was not anyone's intention between about 1860 and the 1890s.

Robert P. Porter, who helped compile the Census of 1880, wrote a book summarizing its findings that celebrated "the march of civilization" from east to west with no misgivings. In writing about Arizona he devoted only one sentence to the Grand Canyon, and of the Colorado River he remarked that although it is the largest in the arid West, it "can not be used to serve any large quantity of land, owning to its bluff, high banks, and cañon walls, and its slight fall where it flows through open country." Using other streams to the maximum, however, Porter estimated that 3.5 million acres could be irrigated in Arizona, roughly the equivalent of an area 100 miles long and 50 miles wide. He also mused on the human history of the region, where, he reported, "are found abundant signs of a former dense civilization, a civilization far in advance of that of the Indians proper now inhabiting those regions, and resembling in many essential features that of the Moquis and Pueblos." The similarity was "so striking as to lead to the irresistible conclusion that the latter peoples are but the remnant of a large and powerful race that once covered with a dense population this great region." Porter then described the ruins of large towns, which he argued predated the cliff dwellings, which were the last resort of a people under attack. The conclusion was clear, "The Navajoes, Apaches, and Utes obtained this country by conquest from this people, and can show no other title."[24] Thus Porter offered a rationale for conquest other than Wheeler's utilitarianism: white Americans were but dispossessing the despoilers of a lost civilization, who had no ancient right to the land. If the rationales differed, the conclusion seemed always to be that white exploration, appreciation, and settlement of the land were inevitable and desirable.

After the public sensation created by his first exploration, Powell immediately went to Washington to raise money for a second expedition, which took place in 1871 through 1872. Much of the valuable scientific work came out of this later trip down the river, including maps of the region and considerable exploration of side canyons. As William H. Goetzmann has noted, Powell later "telescoped his labors of 1869–73 into one exciting report, which made it seem as if his two expeditions down the Colorado had actually been one great and all-encompassing voyage of discovery."[25] Furthermore, he waited more than two decades before composing this book, and by the time of its publication in 1895, Powell's expeditions had transformed into a journey to discover the Grand Canyon. In this version of events, the book's fifteen chapters culminate in a description of the Grand Canyon that is still a useful introduction to the phenomenology of the place, though it can mislead the reader about what the region meant to the first explorers.[26]

The diary of photographer Jack Hillers is more helpful in furthering an understanding of the early river trip, but it was only published in 1972. Reading his laconic entries reminds the reader repeatedly that while Powell was the first systematically to explore the Colorado River, in the same years many others were searching its banks for gold and minerals. Hillers wrote, for example, that on October 6, 1871, near the Crossing of the Fathers, they ran into "two prospectors" who "were anxious to reach some cañon where the water ran swift. They said that they had come all along up the river from the Virgin [River]. Found gold everywhere, averaging from 3–5 dollars a day." The following week, Powell hired another miner to take rations downriver for his expedition "to the mouth of the Pa Weep." Ten days later, Hillers was exploring away from the river and camped with another surveying party and "had a pleasant time."[27] As such encounters suggest, the area was being crisscrossed by prospectors and surveyors, and already it was sparsely settled by Mormons. As early as 1866, the Hualapai Mountains of northern Arizona had been extensively explored, and more than twenty-five hundred mining locations discovered and claimed.[28] Hillers did not have a modern tourist's sensibility. At no point in his diary does he sound disappointed to find that someone else has arrived before him. On the contrary, in the difficult situation of taking wooden boats down an uncharted river full of rocky rapids, these encounters were welcome. Four decades later, the Kolb brothers had the same attitude on their expedition, as they gladly hailed any trapper, prospector, or dredging party sucking up loose gravel from the river bottom that they happened to encounter. In short, the Colorado River valley long was conceived of as both a resource to be exploited and a tourist site.

This attitude suffused an 1882 publication with the revealing title *From River to Sea: A Tourists' and Miners' Guide from the Missouri River to the Pacific Ocean,* which offered alternating chapters on promising mining claims and the natural wonders of the West. The volume included a detailed chapter on the laws of mining immediately after one on sport fishing.[29] A similar attitude toward the region was evident in the second full-fledged expedition down the Colorado River, led by Robert Stanton in 1889 and 1890, with the purpose of surveying several hundred miles of the river as a potential route for the Denver, Colorado Canyon and Pacific Railroad. The idea of a railroad along the Colorado seemed plausible enough in the abstract. A line connecting Grand Junction, Colorado, to the west coast would take advantage of a water grade by going down through the canyons to southwestern California, before leaving the river somewhere between Needles and Las Vegas to cut across to the Pacific.

An engineer, Stanton kept meticulous notes of his explorations, filling more than one thousand pages that were never published in his lifetime. "The engineer's goal was not adventure. It was business . . . Stanton's objective was to find a way to open the canyons to commercial traffic."[30] To this end, he ran a continuous transit line for more than 350 miles along the Colorado; sketched contour topography; and made detailed maps, stopping this activity at Glen Canyon only because funds were low. Below that point, only preliminary work was done. Stanton's party also took more than 1,100 photographs, most of them in duplicate. According to these surveys and studies, construction of a railroad line was technologically feasible. It would not have required many tunnels, but a great deal of blasting to create stone benches for the tracks. The resulting railroad could have run at a gentle grade through spectacular scenery. It could have been free from ice and snow, because of its low elevation, and competed effectively with other lines that ran through high mountain passes. Such a line could have carried Colorado coal to California, where fuel was expensive, arriving mostly from Australia or British Columbia. Such a line could have appealed to tourists, as well as make extensive mining operations feasible all along the Colorado and its tributaries. Gold placer deposits had been found in many places, including several claimed by the expedition itself. It seemed only logical to suppose that in such a vast system of canyons there would also be copper, silver, lead, and other valuable products. Indeed, at the Chicago Columbian Exposition, some particularly rich copper ore from the Grand Canyon was displayed.[31]

Stanton's diary reveals a mind that saw no contradiction between the grandeur of the scenery and railway development. One day he writes that "the views become more grand. The marble mountains come near the river and rise towering over it . . . The bright colors are entirely wanting, but the nearness of the peaks to the river gives one a more correct idea of the greatness of the *whole canyon*." The next day, Stanton evaluates the scene with an engineering aesthetic: "Beautiful railroad bench on granite under the sandstone all around this left turn, 1–40 foot opening, 1–10 foot opening, or if the grade will suit, a beautiful line on flat on top of the sandstone."[32] A rock formation could be both grand and inspiring and a "beautiful railroad bench."

Stanton's detailed calculations and visions never attracted enough investors, although he spent much of his life advocating the railroad. At one point, he hoped to make the canyon into a more lucrative market through an ambitious scheme to dredge for gold in Glen Canyon. For this project he found financial backing, staked 145 claims, and brought in ten tons of drilling equipment. Initial work indicated that there were half a billion cubic yards of gravel with gold content at 25 cents a yard. More equipment was brought in at great cost, and in 1901 the dredge began to operate, producing only a few ounces of gold after a month of operation, when the whole scheme was abandoned.[33]

If mining, especially gold and silver, preoccupied early settlers, prospectors, and surveyors, the possibility of using the Colorado River for irrigation soon became a regional obsession. Homesteading the desert through irrigation seemed a logical extension of Manifest Destiny. In the early 1870s both Wheeler and Powell were advocating extensive use of the river for this purpose. In 1875 Lieutenant Eric Bergland, another West Point graduate, had prepared a detailed investigation into "the feasibility of the diversion of the Colorado River of the West from its present channel, for the purposes of irrigation."[34] Bergland focused on the desert area of southeastern California west of the river, near the Mexican border, which would be renamed the Imperial Valley by the promoters who eventually irrigated the area a quarter century later. Smaller-scale irrigation had been practiced on virtually all of the Colorado's tributaries, first by Native Americans and later by Spanish and Anglo-American pioneers. Near present-day Phoenix, Arizona, the Hohokam tribe built an extensive canal system focused on the Gila and Salt Rivers that was in use for roughly one thousand years.[35] Five hundred years later, the Pima tribe was redeveloping the irrigation system, just as the Anglo-Americans began to settle the region. Native Americans lost most of the conflicts over water in the late nineteenth century, as irrigation became the foundation of early regional growth. On the Nevada-California side of the river, there were few tributaries to exploit. Yet the land was suited to agriculture, especially as one reached the southeast end of California, for here most of the soil was part of a vast Colorado delta. This was the area Bergland had wanted to irrigate, but in 1875 no private developer possessed the resources, and the federal government was not ready to become involved.

Elsewhere in California, irrigation developed rapidly, however. The first combination irrigation-electrification project came already in the 1880s, when a self-educated engineer, George Chaffey, created the communities of Etiwanda and Ontario. The experience gained in such projects led to more grandiose designs. In the 1890s, Chaffey showed investors how to divert the Colorado into an old channel near Yuma and then run the water through a loop into Mexico and back, to irrigate a portion of the Sonoran Desert they had renamed "The Imperial Valley of California." Even before the water began to flow in 1901, the first settlers began to arrive, many coming from Arizona's Salt River irrigation area.[36] However, in the following decade the Colorado River proved too powerful for the floodgates constructed; it deserted its main channel, rampaged over the region, buried the settlers' claims, and flowed into a desert sink called the Salton Sea. Local investors had insufficient resources to deal with the catastrophe and called on the railroads and the federal government for help.[37]

The attempt at economic development of the arid regions of the United States was not long confined to unassisted private enterprise. Congress continually sought ways to stimulate settlement and development of the lower Colorado Basin, at first through laws to encourage railroad construction and homesteading. These techniques had proven their utility in the rapid development of the Great Plains, assisted as well by the short-lived belief that "rain follows the plow." Homesteading the desert clearly required water, however, and in 1902 Congress established "the Reclamation Fund, to be used in the examination and survey for and the construction and maintenance of irrigation works for the storage, diversion, and development of waters for the reclamation of arid and semiarid lands" under the direction of the secretary of the interior.[38] The scope was broad, including dam construction, canals, irrigation, and artesian wells, and the clear goal was economic development. Worster, in *Rivers of Empire,* sees the rise of the Bureau of Reclamation as part of a larger process in the centralization of economic and political power.[39] Irrigation systems, because of their high cost and complex nature, require a powerful central management and tend to concentrate economic resources in a few hands. While many enthusiasts for western irrigation saw it as a way to project Thomas Jefferson's nation of small farmers into a new geographic context, western water development usually did not sustain a new kind of yeoman farmer. Rather, it resulted in either large landholders or tightly knit groups, such as the Mormon towns. The startup costs of irrigation farming were high, the land accessible to water was limited, and linkages between political influence and economic advantage were close.

Although Congress originally established a maximum of 160 acres that any individual could irrigate from publicly supplied water, the Bureau of Reclamation's interpretations of this limit were so varied and relaxed that in practice individuals created much larger landholdings, irrigated at subsidized prices. A single family could in fact subsist on less than 160 acres of land that always had both water and sunshine. In the 1890s Powell advocated a maximum of 80 acres and calculated that there was room for 1,250,000 families on 1 million acres of irrigated land.[40] Something resembling this vision was realized at a few sites, for example, on the Snake River in Idaho, where in 1920 the Bureau of Reclamation completed a hydroelectric dam and a 630-mile canal system. There, 121,000 acres of formerly arid land produced alfalfa, potatoes, sugar beets, and garden crops and made it possible to build up a dairy industry. Furthermore, the dam supplied the farmers with light and power, at a time when 90 percent of American farmers still did not have electricity.[41] But such tightly knit communities of small farmers seldom resulted from the bureau's activities. In most cases, water and wealth were concentrated in a few hands.

In practice, taxpayers funded expensive dams, while a small elite received inexpensive water. This result flouted Congress's original intention, and a good deal of the history of western water is the story of conflict over this 160-acre limit. Fundamentally, however, both sides in the argument agreed that western rivers ought to be developed; they disagreed only on how the benefits ought to be distributed. Both sides wanted to transform the desert into farmland. The fundamental premise was stated baldly on the cover of the Department of the Interior's report *The Colorado River: A Comprehensive Report on the Development of the Water Resources of the Colorado River Basin for Irrigation, Power Production, and Other Beneficial Uses in Arizona, California, Colorado, Nevada, New Mexico, Utah, and Wyoming. "A natural menace becomes a natural resource"*—these words on the title page were repeated in the foreword, which began, "Yesterday the Colorado River was a natural menace. Unharnessed, it tore through deserts, flooded fields, and ravaged villages. It drained water from the mountains and plains, rushed it through sun-baked thirsty lands, and dumped it into the Pacific Ocean—a treasure lost forever. Man was on the defensive. He sat helplessly by to watch the Colorado River waste itself, or attempted in vain to halt its destruction." The Colorado was represented as a thief on the rampage, misusing water in floods, denying it to "thirsty lands," and then throwing its treasure away. In this drama, the river robbed the land, and humans were helplessly "on the defensive." The river was also a wastrel that "wastes itself." Presenting the Colorado as a being whose actions are misguided suggested the possibility of reform, and the foreword continued, "Today this mighty river is recognized as a national resource. It is a life giver, a powerful producer, a great constructive force. Although only partly harnessed by Boulder Dam and other ingenious structures, the Colorado River is doing a gigantic job." Like a wild horse, it has been harnessed: "Its water is providing opportunities for many new homes and for the growing of crops that help to feed this nation and the world. Its power is lighting homes and cities and turning the wheels of industry. Its destructive floods are being reduced. Its muddy waters are being cleared for irrigation and other uses." Rationality has controlled and clarified the stream and made it work for America. Engineering planning is presented as the ideal mode of social development, a vision that by 1946 had already produced the Tennessee Valley Authority and its network of hydroelectric dams.

The report celebrated the final result: human domination of nature:

> Tomorrow the Colorado River will be utilized to the very last drop. Its water will convert thousands of additional acres of sagebrush desert to flourishing farms and beautiful homes for servicemen, industrial workers, and native farmers who seek to build permanently in the West. Its terrifying energy will be harnessed completely, to do an even bigger job in building the bulwarks for peace. Here is a job so great in its possibili-

> ties that only a nation of free people have the vision to know that it can be done and that it must be done. The Colorado River is their heritage.[42]

The river's meaning is reduced to use value. Its water exists primarily for human beings, and its energy is to be completely subjugated. Furthermore, the end of World War II is invoked in the image of beautiful farms and homes for servicemen, while the development of the basin becomes part of creating "bulwarks for peace." The text goes still further, suggesting that only a free democratic people could sustain such a vision and see it as a historical necessity. The Colorado River thus functions not only as a wild thing to be subdued or a wasted energy source to be rationally developed. The complete development of the Colorado is a part of America's historical fate, an actualization of its Manifest Destiny. A free people will see that they must dominate nature. To realize their freedom, they must recognize, paradoxically, that they have no choice. The river is "theirs" not only in the sense that it is physically located (primarily) within the United States, but also in a teleological sense. Its domination is necessary as part of the larger historical narrative of American expansionism. The dam-building and irrigation projects are inevitable expressions of national growth. While the rest of the official report does not maintain this lofty rhetorical tone, all of its statistics, maps, and projections are predicated on the promotion and continuation of migration into this region.

Similar plans had been expressed in one way or another by every previous secretary of the Interior, perhaps never so succinctly as by John W. Noble, who in 1893 declared, "All through that region, much of which is now arid and not populated, will be a population as dense as the Aztecs ever had in their palmiest days in Mexico and Central America. Irrigation is the magic wand which is to bring about these great changes."[43] By the time of the 1946 report, earlier population growth could be adduced as evidence of potential future expansion. Each dam justified the next. Controlling nature was thought to be the logical destiny of a free people.

Despite the developers' vision of homes for returning servicemen, irrigated farming, and the coming of industry, the lower Colorado Basin would in fact develop quite differently. The largest city near the river turned out to be Las Vegas, and the largest consumer of its water turned out to be outside the watershed altogether: Los Angeles. The Jeffersonian ideal of a rural nation based on farmers was used to justify dam building and homesteading by irrigation, with an ironic result. Instead of small farms, agribusiness controlled much of the water. Instead of independent farmers, these large enterprises were like factories in the field, which demanded cheap labor, much of which came from Mexico. Instead of dispersing people and maintaining a rural nation, the dams encouraged population concentration and urban growth.

Nevertheless, the brochure handed out to tourists in 1996 still reads like a precis of the Department of Interior's 1946 report, *The Colorado River.* The dam is presented as "an engineering wonder" with "multipurpose benefits." It concludes with a virtual verbatim quotation of that earlier report: "Hoover Dam changed the Colorado River from a natural menace to a national resource, strengthening the economy of the Southwest and the Nation."[44] It had paid off the original construction cost, plus 3 percent interest, and its seventeen generators were being "uprated" so that they could produce a peak load of almost 50 percent more power. Utilitarian millennialism remains the gospel of the tour guides and the publications available in the gift shop. To the Bureau of Reclamation, the Colorado River was not the kind of heritage to be preserved in a museum or kept intact. Yet this is precisely what a growing number of Americans began to demand in the postwar era.

Contemporary tourists, like the explorers of the 1870s, shift aesthetics effortlessly when moving from one site to the next. When visiting the Grand Canyon, most families appear imbued with the values of preservationism and the ideology of America as nature, but when the same families visit Hoover Dam, usually within a day of seeing the canyon, they seem immediately able to adopt the utilitarian, developmental ideal. Americans of the 1990s are embracing once again the contradictions to be found in Powell, King, Wheeler, Stanton, and other nineteenth-century ancestors. As Jacob Wamberg argues elsewhere in this volume, the appreciation for landscape emerged along with the revalorization of labor. Similarly, opportunities for resource development have usually been seen as not merely utilitarian but also as having a redemptive quality. They are the uplifting labor of industrialism.

The same sensibility that appreciates the sublime view, with its linear perspective into a morally uplifting infinity, implicitly embraces a topographical approach to space itself. In this neutral space, the world is raw material for the gaze to play over, whether it seeks the sublime or the utilitarian.[45] An individual could expand the self through an encounter with the sublime or through the sense of power gained from the exploitation of resources. This Cartesian sense of space, whether expressed in terms of development or in terms of the sublime, includes Native Americans as part of the foreground. They even became a central part of the visual economy, after they were understood by the Santa Fe Railroad and others as an exotic resource that could be exploited.[46] The Indian formed an essential counterpoint to modernity.

Like dam building along the Colorado, the sheer power cascading over the brink at Niagara Falls inspired many to imagine harnessing its waters. To many progressives, the extensive "hydroelectric development of Niagara in the late 1890s seemed to be a sort of capstone on humanity's victory over nature." It seemed possible that "the development of Niagara represented the beginning of a new, totally human order."[47] The utopian novel by King Camp Gillette, *The Human Drift,* announced that one great city called Metropolis, with 60 million people, could focus the energies of Niagara in a perfect society, run by a World Corporation. The falls were to power an enormous factory district (that would be as automated as possible), encircled by a huge residential area that would house the entire population of the United States. Gillette imagined the rest of the country "swept clean of cities, towns, villages, farmhouses, country roads, fences, and railroads" replaced by "one perfect, beautiful and rational central city."[48] The imperfect settlement of America is imaginatively wiped out and a new structure put in its place. Niagara Falls would be put at the service of industry, but in exchange, all the rest of America would cease to be exploited.

This proposal resembles the demands of environmentalists who want more wilderness areas. Such ideas implicitly suggest their polar opposite, civilization concentrated somewhere else. Gillette's utopia articulated a fundamental contradiction within the impulse toward industrial progress, akin to the dreams of an irrigated utopia in the West. His fantasy acknowledged that the cities and towns already constructed were a blot on the landscape that was best removed by utterly subordinating Niagara to industrialization.

Other natural wonders have also called out the impulse to development. Yellowstone's thermal springs and geysers stimulated fantasies of commercial development that were successfully quashed by those who lobbied to make it a national park. The Colorado River inside the

Grand Canyon was repeatedly proposed as a hydroelectric dam site, and legislation to this effect was introduced in Congress as late as the middle 1960s, raising a storm of protest from the environmental movement.[49] At such moments sublimity may appear to be at the furthest remove from the utilitarianism, but in America the technological is often intertwined with the sublime.[50] The vast and powerful objects of nature seem to offer themselves up for both ecstatic contemplation and improvement. Thus Frank Waters could write a book on the Colorado River that both admired the sublimity of the Grand Canyon and enthused over the newly constructed Hoover Dam, which he called "a fabulous unearthly dream. A visual symphony written in steel and concrete."[51] This apparent inconsistency recurs from the explorers of the nineteenth century down to the middle 1960s for almost all writers, revealing that wilderness appreciation and utilitarian development long remained inseparable from the American social construction of landscape.

NOTES

1. See David E. Nye, "Die Niagara-Falls, der Grand Canyon und das Erhabene," *Zeitschrift für Semiotik* 19, nos. 1–2 (1996): 7–20, and David E. Nye, "De-Realizing the Grand Canyon," in *Emotion and Postmodernism,* ed. Gerhard Hoffmann and Alfred Hornung (Heidelberg: C. Winter, 1997), 75–94.

2. "America's Magnificent Seven," *U.S. News and World Report,* 21 April 1975, 56–57.

3. J. Donald Hughes, *In the House of Stone and Light: A Human History of the Grand Canyon* (Flagstaff: Grand Canyon Natural History Association, 1978), 28.

4. Joseph C. Ives, *Report upon the Colorado River of the West* (Washington, D.C.: U.S. Government Printing Office, 1861), 131. Copy, Leeds University Library.

5. Ibid., 109–110, 123.

6. Ibid., "Hydrographic Report," 1–14.

7. Mark Twain, *Roughing It* in *The Works of Mark Twain,* vol. 2, ed. Frederick Anderson and Robert H. Hirst (Berkeley: University of California Press, 1974).

8. Doris Ostrander Dawdy, *George Montague Wheeler: The Man and the Myth* (Athens: Swallow Press / Ohio University Press, 1993), 2, 7.

9. Cited in Alan Trachtenberg, *Reading American Photographs: Images as History, Matthew Brady to Walker Evans* (New York: Hill and Wang, 1989), 146.

10. Quoted in Dawdy, *George Montague Wheeler,* 9.

11. Michael Smith, *Pacific Vistas: California Scientists and the Environment, 1850–1915* (New Haven, Conn.: Yale University Press, 1987), 84.

12. John Wesley Powell, *The Exploration of the Colorado River and Its Canyons* (1895; New York: Dover Publications, 1961).

13. See Donald Worster, *Rivers of Empire: Water, Aridity, and the Growth of the American West* (New York: Pantheon, 1985), 138–139.

14. Donald Worster, "Comment: A Response to 'John Wesley Powell and the Unmaking of the West,' " *Environmental History* 2 (April 1997): 218.

15. John W. Powell, *Exploration of the Colorado River of the West and Its Tributaries, Explored in 1869, 1870, 1871 and 1872* (Washington, D.C.: U.S. Government Printing Office, 1875), 113, 161–167, passim.

16. Smith, *Pacific Vistas,* 74.

17. See Elizabeth McKinsey, *Niagara Falls: Icon of the American Sublime* (Cambridge: Cambridge University Press, 1985), 37–125.

18. Stephen J. Pyne, *Dutton's Point: An Intellectual History of the Grand Canyon,* Monograph 5 (Flagstaff: Grand Canyon Natural History Association), 34.

19. Joni Louise Kinsey, *Thomas Moran and the Surveying of the American West* (Washington, D.C.: Smithsonian Institution Press, 1992), 111–116.

20. See the thirty-one-page pamphlet, which was frequently reissued, by Charles A. Higgins, *Grand Cañon of the Colorado River, Arizona,* illus. Thomas Moran, H. F. Farny, and F. H. Lungren (Chicago: Passenger Department, Santa Fe Route, 1893).

21. Dawdy, *George Montague Wheeler,* 71.

22. Quoted in Albert Boime, *The Magisterial Gaze: Manifest Destiny and American Landscape Painting, c1830–1865* (Washington, D.C.: Smithsonian Institution Press, 1991), 140, 141.

23. Dawdy, *George Montague Wheeler,* 42.

24. Robert P. Porter, *The West: From the Census of 1880* (Chicago: Rand, McNally, 1882), 10, 456, 460, 462.

25. William H. Goetzmann, foreword to *A Canyon Voyage: The Narrative of the Second Powell Expedition Down the Green-Colorado River from Wyoming, and the Explorations on Land, in the Years 1871 and 1872,* by Frederick S. Dellenbaugh (1908; reprint, New Haven, Conn.: Yale University Press, 1962), xviii.

26. Powell, *Exploration of the Colorado River,* 379–398. To disentangle Powell's journeys of 1869 through 1873 from his evocation of them a generation later, it is only somewhat helpful to read the only other full-length account written by a member of the second expedition, Frederick S. Dellenbaugh, since he did not get around to writing until after Powell's work appeared. When Dellenbaugh's two volumes appeared in 1904 and 1908, fully three decades after the journey, the Grand Canyon had become a celebrated tourist site, and the original journey had already acquired mythic dimensions, especially because virtually no one had ventured down the river since then.

27. Don D. Fowler, ed., *"Photographed All the Best Scenery": Jack Hillers's Diary of the Powell Expeditions, 1871–1875* (Salt Lake City: University of Utah Press, 1972), 85, 87, 89.

28. Dawdy, *George Montague Wheeler,* 34.

29. Chas. S. Gleed, ed., *From River to Sea: A Tourists' and Miners' Guide from the Missouri River to the Pacific Ocean* (Chicago: Rand, McNally, 1882). The Grand Canyon is not mentioned by name, though the canyon country of the Colorado is mentioned in passing, and while there are woodcuts of Yellowstone, Yosemite, Garden of the Gods, and many other western scenes, the Grand Canyon is not depicted. By the end of the next decade this omission would have been unimaginable.

30. Dwight L. Smith and C. Gregory Crampton, eds., *The Colorado River Survey: Robert B. Stanton and the Denver, Colorado Canyon & Pacific Railroad* (Salt Lake City, Utah: Howe Brothers, 1987), 4.

31. I thank Hal Rothman for this information.

32. Smith and Crampton, *Colorado River Survey,* 254–255, emphasis in original.

33. David Lavender, *Colorado River Country* (New York: E. P. Dutton, 1982), 148–161.

34. Dawdy, *George Montague Wheeler,* 74.

35. Specialists disagree about why it was abandoned in the mid-fourteenth century. Theories range from increasing alienation to overpopulation to outside interference to disease. See James L. Wescoat, "Challenging the Desert," in *The Making of the American Landscape,* ed. Michael P. Conzen (London: Routledge, 1994), 190 and passim.

36. Worster, *Rivers of Empire,* 196.

37. A mythology grew up around this irrigation project, which became the subject for two popular novels, Ednah Aiken's *The River* (Indianapolis, Ind.: Bobbs-Merrill, 1914), and Harold Bell Wright's *The Winning of Barbara Worth* (1911; reprint, New York: Grosset & Dunlap, 1966). The latter was made into a silent film in 1926 by Samuel Goldwyn.

38. Language of the original law quoted in U.S. Department of the Interior, *The Colorado River: A Comprehensive Report on the Development of the Water Resources of the Colorado River Basin for Irrigation, Power Production, and Other Beneficial Uses in Arizona, California, Colorado, Nevada, New Mexico, Utah, and Wyoming* (Washington, D.C.: U.S. Government Printing Office, 1946).

39. Worster, *Rivers of Empire,* passim.

40. Ibid., 138–139.

41. David E. Nye, *Electrifying America: Social Meanings of a New Technology* (Cambridge, Mass.: MIT Press, 1990), 300.

42. U.S. Department of the Interior, *The Colorado River,* 25.

43. Quoted in Worster, *Rivers of Empire,* p. xi.

44. Bureau of Reclamation, "Hoover Dam," brochure, 1993.

45. I explore this idea in *Narratives and Spaces: Technology and the Construction of American Culture* (New York: Columbia University Press, 1998).

46. Marta Weigle and Barbara A. Babcock, eds., *The Great Southwest of the Fred Harvey Company and the Santa Fe Railway* (Phoenix, Ariz.: Heard Museum, 1996).

47. Patrick V. McGreevy, *Imagining Niagara: The Meaning and Making of Niagara Falls* (Amherst: University of Massachusetts Press, 1994), 106.

48. Ibid., 134.

49. Roderick Nash, *Wilderness and the American Mind,* 3d ed. (New Haven, Conn.: Yale University Press, 1982), 227–237.

50. David E. Nye, *American Technological Sublime* (Cambridge, Mass.: MIT Press, 1994), 32–76, passim.

51. Frank Waters, *The Colorado* (New York: Rinehart, 1946), 337.

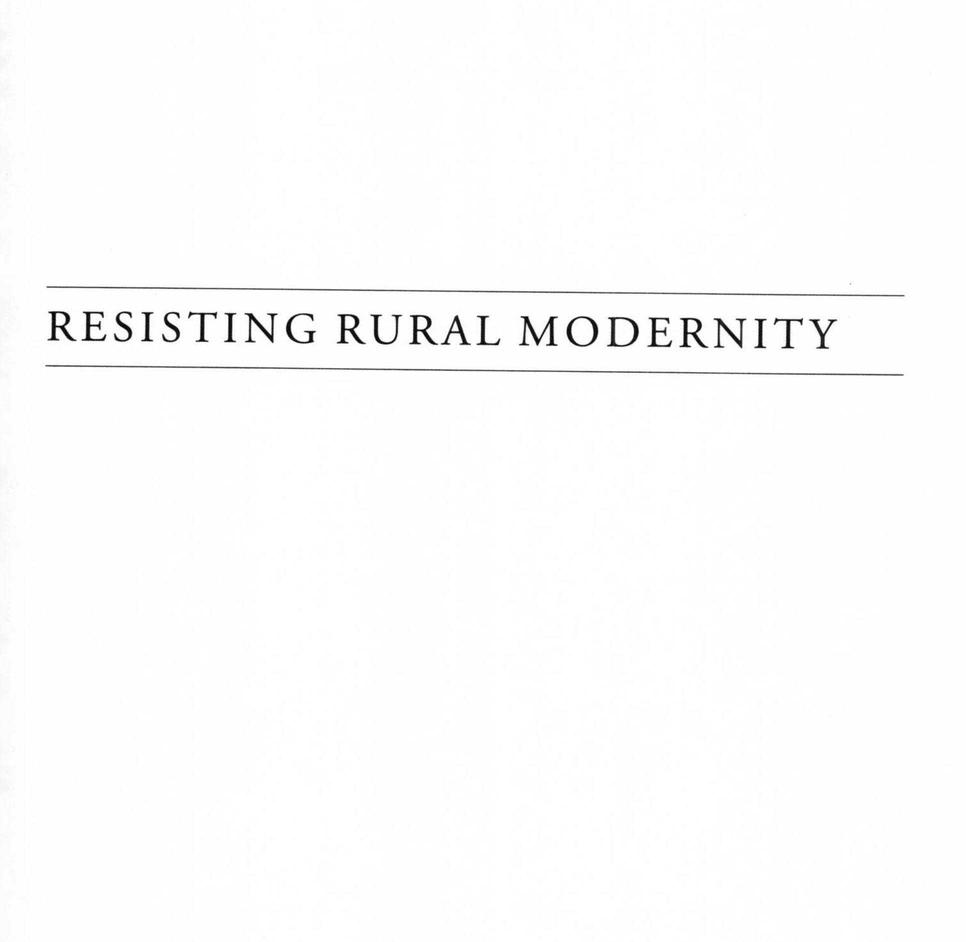

RESISTING RURAL MODERNITY

6

Begrudging Aesthetics for a New South

The Farm Security Administration Photographic Project and Southern Modernization, 1935 to 1943

STUART KIDD

Prevailing cultural representations of the South during the 1930s were deeply contradictory. If the dominant stereotype was that of a "blighted" South of tenants and sharecroppers, the region's premodernity invited an alternative construction that was more comforting in the contexts of the Great Depression and escalating global tensions: namely, that the South was a repository of American tradition and values. No source made a more substantial contribution to this ambiguous cultural imagery than the Farm Security Administration (FSA) photographic project. Under the direction of Roy Stryker, photographers affiliated with the Historical Section produced nearly sixteen thousand images of the South between 1935 and 1943. Widely reproduced in the national media, the FSA photographs offered both "social criticism" and "social solace."[1] Also, they helped promote a general interest in the region, which was registered in the baffled observation of the journalist Jonathan Daniels that visitors to the South during the Depression era preferred to visit a sharecropper's cabin than to behold an azalea.[2]

Of course, there was yet another representational South of the 1930s: that of the "New South." So strong is the case that the decade constitutes a watershed between the South's rural plantation system and the modernized and diversified economy of the "Sunbelt" that, for historians, the question is no longer whether the age was significant, but whether it was the Depression, the New Deal, or World War II that caused the collapse of the old southern order.[3] Generally, however, this dimension is underrepresented in the FSA's photographic file, for, paradoxically, none of the principal forces instrumental in its making was comfortable with the South that was emerging at this historical juncture.

The FSA itself, although committed to rejuvenating depressed rural areas, also sought to maintain traditional patterns of landowning and, despite the increasing momentum of agricultural mechanization, championed the maintenance of the family farm. Regional economic modernization outside the orbit of its own program not only was incompatible with many of the agency's operations but threatened its very raison d'être. Stryker, who effectively renegotiated the Historical Section's remit to expand his photographers' compass to include subjects other than those directly related to either agricultural poverty or federal assistance, was temperamentally averse to modernization. Although Stryker was a social scientist by training and an innovative administrator who sought to transcend the publicity functions of his unit by developing a social scientific research role for it, sentiment and sociology fused in his documentary vision, for he was deeply conscious that the society under scrutiny was being transformed and, often, he believed, not for the better.[4]

Stryker's principal photographers were based in the urban centers of the east and west coasts of the United States. For them, the American heartland was terra incognita. "I lived in an isolated world," recalled Carl Mydans, who was Boston-born and had worked in New York City before joining the Historical Section. "Vaguely there was a great river called the Hudson, and a great bit of land at the other side of the Hudson, but I knew very little about it." The opportunity to travel the country as a government employee was both educational and revelatory. Jack Delano reflected, "I had always been a city boy and it was all great and new and wonderful to me." "The word 'land' meant little to me," he recalled. "Like some exotic animal in the zoo, it was kept in a special place called Fairmount Park, where you might go see it on weekends or holidays."[5] The American heartland became as much as an imaginative space as it had been for the nineteenth-century landscape painters, for whom geographical discovery opened up a range of creative possibilities. For the photographers, the natural and human environments of the South provided a touchstone for the modern world of mass and motor in which they lived and pursued their professional careers. If the photographers' terrain did not possess the natural grandeur of the painters', its inspirational potential was no less.

The photographers' sensibilities were not at variance with either the socioeconomic objectives of their employing agency or the documentary ambitions of the Historical Section's head, for landscapes of premodernity offered creative possibilities while satisfying the FSA's need for effective publicity and Stryker's eagerness to catalogue a vanishing America. While in the South, Stryker's photographers were absorbed by the aesthetics of habitats that were built of natural materials, were on a human scale, and had acquired distinctive appearances due to the eccentricities of construction and repair and the ravages of weather. They stood in stark contrast to the anonymous and uniform brick, mortar, and glass apartment blocks that provided living space for urban Americans. The close-up studies of rickety steps, mud and stick chimneys, rainstained walls, rotting weatherboards and bowed piles of foundation stones not only contributed to the individual characters of the buildings, but also registered lifestyles that were not so much obsolete as in an advanced state of aging. The natural ravages of a sharecropper's cabin invited a respect that a rundown urban tenement could never command.

In their studies of abandoned and derelict mills, farmhouses, barns, and outbuildings, the photographers depicted culture reverting to nature and humans creating their own wilderness. Essentially, the photographers inverted the terms of Frederick Jackson Turner's "frontier thesis." Humans in stasis had returned to a near-primitive state, their presence expressed in vestiges rather than monuments. However, the vision is not that of Erskine Caldwell. These are images of decay rather than of squalor. Decay presented photographers with a range of visual possibilities. Russell Lee's study of a plantation house destroyed by fire in Lutcher, Louisiana, is a landscape of irrationality. The stairs lead nowhere, the columns support nothing, and the walls enclose only vacant space. It is a landscape of mystery and romance, Faulknerian in its resonances of the past. In Ben Shahn's study of the remains of the Gulf State paper mill, a landscape of devastation becomes an engaging reconstitution of related objects. Tragedy provided an opportunity for experimentation with formal perspective.[6]

The photographic record of the Southern Series makes abundantly clear that the old order, which underpinned southern premodern culture, had collapsed. Photographers used plantation house remnants to symbolize the demise of the traditional southern agricultural system, the lone chimney being a recurrent motif. The perspectives contained both sociology and sensibility. As literal narratives they suggested social and economic ill-being. As compositions they engaged with symbolism and myth. The crumbling columns, collapsed verandas, and sagging balconies, sometimes shot in close-up, register a faded grandeur and an architectural character perversely enhanced by decay. Silent and still dinner bells on the lands of abandoned plantations were redundant instruments and material survivals of the doomed agricultural order. A quality that distances the photographers' work from both social scientific documentary and didactic publicity is the sense of loss that much of it conveys.

Degeneration had manifold charm for the photographers, and perhaps herein lies the essence of the contradiction between romance and recrimination in their engagements with the southern rural landscape. While they generally avoided the picturesque and understood their assignments in terms of the political nature of their work, they produced many images that dignified the very conditions the FSA was attempting to eradicate. For them, the landscape of the South was a stimulating antidote to the predictable, anonymous, and modernized cities with which they were associated professionally. Man's traces on the southern landscape, in contrast, were personalized, rooted in history and culturally "authentic." The beleaguered status of these vestiges reinforced their imaginative appeal.

In much FSA photography, the modern world serves as a touchstone or an embedded counterpoint for the rural South. The references are most direct in showing the ingenuity of tenants who used the materials of commercial America in their struggle for survival. Roofs made out of license plates, magazines and newspapers providing insulation and filling cracks between logs and boards, and stoves made out of oil cans all bear poignant testimony to a society holding itself together with the detritus of the civilization that is replacing it. Conversely, in much of the coverage of southern rural industry, it appears that the endogenous forces of regional decline are too strong for the exogenous forces of economic recuperation. In their studies of abandoned sawmills, cigar factories, turpentine stills, cotton gins, and paper mills, the photographers suggest that industry suffered the same fate as the plantations within a shorter time span. The suggestion is that no economic order could thrive in the South and that, in John Collier's words, "a broken past is (a) poor foundation for a functioning future." It was this sense of disestablishment that provided the photographers with interpretative possibilities in the southern region.[7]

The burgeoning industrial environments in the South were less inspirational. Although Stryker had successfully convinced his superiors to allow a more holistic approach to regional photographic documentation, the holdings on the South's principal industries in the Historical Section's file were meager. Underrepresentation of the iron and steel, cotton textile, coal, tobacco, and lumber industries was partly a product of the owners' sensitivity to adverse publicity and their refusal to cooperate with the project. This certainly accounts for the limited coverage of the iron and steel industry. What is visually most significant about the Southern Series' holdings on iron and steel is the distance between the photographer and the industry. There are no images of the production process and few of the workers. For the most part, the industry is represented as a skyline of kilns, smokestacks, and hoists towering over workers' homes. The compositional form was conceived by Walker Evans in 1936. In the best-known image of his series, the steelworkers' homes and their outdoor privies run in parallel lines across undulating land like curved arteries from the steelmill in the background. Subsequently, Arthur Rothstein and Marion Post Wolcott would employ Evans's perspective more crudely, sharpening the contrast between the domestic hearth and the industrial furnace by foregrounding workers' cabins against backdrops of towering chimneys. However, imitation was not only a gesture of respect for Evans's artistry.[8] It signifies the fact that the company town and its mills were off-limits to Stryker's photographers.

Lack of cooperation by textile mill owners also forced Stryker to shelve a project on the industry in 1939. However, this alone does not account for the dearth of images in the file. Partly, it was a result of the reluctance of Post Wolcott, Stryker's principal photographer in the South during 1939 and 1940, to undertake industrial assignments. She evaded several on the grounds that the season was unsuitable. In July 1939, she decided not to photograph the mill towns of the Piedmont in a midsummer setting. "It's absolutely the wrong time of year," she wrote, "much too pretty and everything covered with much trees and shrubbery and gardens, etc." A year later, she offered similar reasons to justify her reluctance to do work in Kentucky's coalfields. It was not the violent reputation of Harlan County that deterred her, but the "goddamn shrubs and trees and weeds [which] cover everything up now" and which detracted from the industry's "true character." Post Wolcott preferred a winter assignment before the "leaves pop."[9] Point of view as much as accessibility and formal perspective accounts for the photographers' distance from southern industry. Post Wolcott's own stipulations about the appropriate conditions for industrial photography reveal feelings about southern industry that are also implied in an unexceptional exterior study of an Atlanta textile mill, one of the few taken before 1941. She highlighted its only salient feature in a short caption: "A high fence topped with barbed wire around a cotton mill." The meaning was clear to Arthur Raper, who used the photograph in *Sharecroppers All* accompanied by his own commentary: "Cotton millowners know their erstwhile rural workers are not dependably docile."[10] Generally, however, landscape was not a satisfactory medium for photographers who had no empathy for their subject.

Southern modernization was most apparent in the mechanization of agriculture, which had direct implications for the work of the FSA—some of the best-known images of the photographic project engage with the subject.[11] However, only in 1939 did the agency give belated and serious attention to the mechanization phenomenon, and this was due to the publication of John Steinbeck's novel *The Grapes of Wrath* and its documentary counterparts, Paul Taylor and Dorothea Lange's *An American Exodus* and Carey McWilliams's *Factories in the Field,* which raised public interest in the nation's changing agricultural order. Lange, who was also a photographer in the Historical Section, had actively but unsuccessfully campaigned for the FSA to give more consideration to the social and economic implications of technological change in southern agriculture after she had witnessed the displacement of tenants from the Dibble Plantation in Arkansas during January 1936 and their evacuation to the Delta Cooperative at Hillhouse, Mississippi, which was supported by the Southern Tenant Farmers' Union. During her tour of the South in 1937, Lange came to appreciate that such incidents were not isolated examples of exploitation but aspects of a process that was having a profound effect on agriculture and undermining the effective delivery of the FSA's own programs on behalf of small farmers. Lange's awareness of technological change was sharpened by her marriage to Taylor, an economist from Berkeley and the instigator of the Resettlement Administration's migrant assistance program in California, who accompanied her on her summer tour. In the rich cotton lands of the Texas Panhandle and the Mississippi Delta, they witnessed what the historian Pete Daniel has referred to as "the southern enclosure," in which land was being consolidated and tenants were being reduced to the status of day laborers. "It's a force that can hardly be checked even if we would," wrote Taylor. "The cotton sections of the South are in for their greatest revolution since the Civil War, and it's already under way." The social consequences were of particular concern to Lange. After watching fleets of trucks at first light transporting fieldhands from Memphis to plantations in neighboring Arkansas, she reflected, "All of them laborers: hundreds and hundreds, men and women . . . You couldn't find a better example of exploitation and its consequences—in this nation—than the changes in plantation life with this tidal wave of tractors."[12]

In the Texas Panhandle, Lange exploited the opportunities offered by the vast and empty spaces of the Great Plains to distill the changes imposed by power farming. Her images of abandoned farmhouses, marooned in swathes of mechanically plowed furrows in Childress and Hall Counties symbolized a superseded agricultural order and registered the extinction of husbandry by agribusiness. "Tractored Out," the best-known photograph of this series, features a tenant's farmhouse standing illogically intact in the path of contoured furrows that signify a new productive system and, in their uniform lines and patterns, the conquest of land by technology. The care Lange took to historicize her images made her landscapes poignant as well as pleasurable. The poet Archibald MacLeish used one of them in his illustrated poem *Land of the Free* to reflect on the precariousness and rootlessness of American life. MacLeish wrote, "We've got the public highway when the tractors / Crowd the hoes off till the houses sit there / Empty and left as though the floods had hit them."[13]

The drama of Lange's Texas landscapes could not be duplicated in places where space was more confined and patterns of land use more traditional. In the Mississippi Delta, fields plowed mechanically were not visually distinct from those cultivated by mule and plow. In Mississippi, Lange relied less on landscapes than on recording the operations of tractors and their drivers to convey the impact of technology on agriculture. Moreover, she and Taylor may have felt that even the suggestive Texas studies did not adequately convey the turbulence that power farming created. When they illustrated the changes wrought by mechanization for Taylor's employer, Thomas Blaisdell of the Social Security Board, they chose images of displaced tenants accompanied by captioned testimony rather than landscapes to buttress their argument. The "seven Texans" depicted in these images were all young, displaced farmers who supported twenty-nine people and eked out a living by working for the Works Progress Administration or as agricultural laborers.[14] Indeed, toward the end of her career with the Historical Section, Lange became skeptical about the capacity of photography to record "processes," believing that motion pictures were a more effective medium. Assigned to the University of North Carolina in 1939 and Howard W. Odum's "Sub-Regional Laboratory Project," she criticized a shooting script because it contained "too much emphasis on the economic set-up, not enough on the people at the center of it."[15]

Lange's Texas landscapes were exceptional within the FSA's photographic file. Once Stryker began to assign his photographers to the subject of mechanization during 1939 and 1940, he did not choose to emphasize the relationships between land, technology, and people. Rather, he directed his crew to record the technical detail and operation of farm machinery and, in particular, the impact of technology on the rural population. The Historical Section, like its parent agency, negotiated the new rural order more effectively through its coverage of the displaced farmers and migrant laborers who were products of agricultural modernization. Memorable series on migrant labor were produced by Post Wolcott in Florida, Jack Delano in North Carolina and Virginia, and Lee in Louisiana. From the agency's perspective, migrants were landless farmers who required help from the FSA's migrant assistance program. However, there was also a representational logic to engaging with the emergent order as a condition rather than as a process. Images of oppressed migrants, often reinforced by their captioned testimonies, presented an unequivocal indictment of the social effects of modernization and, unlike Lange's Texas landscapes, did not obscure point of view by conflating it with aesthetics.

As creative intellectuals, photographers preferred to engage with the "problem" South rather than any "modernized" South, either in being or which the FSA sought to create. Routine "project work" was resented by most staff photographers. Evans photographed only one resettlement community project, at Briar Patch Farms, Georgia, and claimed to have refused to do any more. Shahn found the "neat little rows of houses" in resettlement communities "impossible to photograph," and he told Richard Doud in the 1960s that "this wasn't my idea of something to photograph at all." Even a faithful journeywoman like Post Wolcott objected to her use as "the project photographer in chief."[16]

During his own travels in the South, the journalist Jonathan Daniels attributed the projects' lack of character to their standardized production line appearance. The Tennessee Valley Authority's (TVA's) town of Norris was so new that it did not have a burial ground, which led Daniels to reflect that "a town without a cemetery [is] a town created without pain."[17] Resettlement community projects offered the photographers few creative opportunities. Their new, neat, functional structures may have captivated a formalist dedicated to precisionism, but in the eyes of Stryker's photographers such dull uniformity did not make for inspiring documentary photography. Barely established, often in a denuded landscape still showing the signs of construction activity, these were truly "half-born" communities. Regional officer George Wolf also recognized that the landscapes of the Deep South, in particular, lacked visual appeal. Responding to a request for advice on how to service a publicity drive on agricultural cooperatives late in 1939, Wolf singled out Clover Bend, a farm community in Lawrence County, Arkansas, for its "excellent pictorial possibilities." "It is one of the few of our projects which has trees," Wolf advised. "It also has a river."[18]

In contrast, the traditional small towns of the South intrigued Stryker and his photographers alike, for, as Information Division staff writer Mark Adams wrote of Lee's work in San Augustine, Texas, "The thing which makes a small town different from a three-hundred family development in the Bronx—the thing that makes it distinctive, and interesting, and whole, is its tradition."[19] Reinforcing the appeal of the South's hamlets, as refracted in so many of the Historical Section's images, was their beleaguered condition. Homogenizing cultural agencies such as radio and the cinema were weakening their parochialism, while centripetal economic influences such as transportation, chain stores, and national sales organizations were transforming them into satellites of metropolitan America. The author Sherwood Anderson believed that the process had already reached an advanced stage. Anderson secured the Historical Section's cooperation to provide illustrations for *Home Town* (1940), which was intended, in part, to be a requiem for the traditional rural village. "I want to bring out how the radio and the movies have changed life in the towns," Anderson advised Stryker. "It has pretty much broken up the little assembly of citizens that used to gather at the back of the drug store, the harness shop, or some other place."[20]

The romance of the southern country town was especially compelling in 1939 and 1940 as the United States embarked on its wartime preparedness campaign and Stryker's unit made its major effort on photographing small towns. In his reflections about Evans's study of the railroad station at Edwards, Mississippi, Stryker wistfully counterpointed two technological eras in a lament on the passing of the small-town community.

> The empty station platform, the station thermometer, the idle baggage carts, the quiet stores, the people talking together, and beyond them, the weatherbeaten houses where they lived, all this reminded me of the town where I had grown up. I would look at pictures like that and long for a time when the world was safer and more peaceful. I'd think back to the days before radio and television when all there was to do was go down to the tracks and watch the flyer go through. That was the nostalgic way in which those town pictures hit me.[21]

However, Stryker did not just require from his photographers rustic detail or keynote pictures that caught the mood of the mythical American village. The water trough, blacksmith's shop, horse and buggy, and oxcart, staples of the small-town landscape in the South, were not solely objects of antiquarian interest, for Stryker's photographers revealed how they coexisted with the service station and automobile of the modern era. In effect, embedded within many of the Historical Section's images of the small towns were narratives of contesting cultures.

Stryker recalled that his photographers were "fascinated" by the automobile and its proliferation throughout America. Even Lange, the photographer most accustomed to the rural poor's possession of automobiles through her work with the Dust Bowl migrants of the Southwest, was impressed by how far down the social order of the South the automobile had reached. After photographing a tobacco sharecropper repairing an inner tube in Douglas, Georgia, she observed, "Fords are beginning to appear as a means of transportation for sharecroppers. They can take tobacco to town and bring supplies back, and also, take workers to Florida for seasonal fruit picking jobs." The southern town provided photographers with a host of antinomial references: horses sharing the highways and main streets with automobiles and trucks, lines of cars parked beside hitching posts, or a camping trailer drawn up alongside a faded mansion. The photographers used these visual juxtapositions so often that they became clichés within the photographic file. Significantly, the first item on a small-town shooting script produced by Stryker in 1939 requested pictures featuring "cars and horses and buggies."[22]

In contrast to their interest in the South's traditional market centers, Stryker's photographers displayed little enthusiasm for the defense boom towns on which they worked as part of the wartime preparedness drive. They photographed green pine cafes and bunkhouses hastily constructed alongside mud-rutted streets and lining the roads leading into town, stores being demolished or refurbished to cater to high-volume trade, and main streets excavated to provide the necessary services for an expanded population. These small towns acquired a provisional character expressed in the marquees that served as motion picture theaters and the shacks built by workers with materials pilfered from construction sites.

Portrayals of defense workers living in trailers, abandoned buses, and streetcars; sleeping in cars; and sharing their encampments with swine were sometimes as shocking as those produced by Post Wolcott in Florida's migrant labor camps. Delano described the settlements around Fayetteville, North Carolina, as "one hell of a mess," while Collier advised Stryker that conditions in Childersburg, Alabama, were "sordid and spectacular."[23] Childersburg, with its temporary housing, overcrowded schools, uncollected garbage, and pit privies dug a block from the town's main crossroads, exhibited some of the worst features of urban blight, according to George C. Stoney. However, the prevailing characteristic referred to by the photographers in their correspondence, and the most appropriate metaphor to describe the changes occurring to this small town, was dust. "Childersburg and its environs are so unpleasant to those who must spend their week days breathing its dust," Stoney wrote. His sentiments were echoed by Collier, who worked in Childersburg in the summer of 1942:

> Dust-smeared men drive construction trucks furiously down unsurfaced roads—red earth rises in a cloud—to settle gently in the windless air, on

> the leaves of the trees crowding the river—on the leaves of the young corn—on the stalks of young bullrushes—on the sleek tops of trailers, across the warped planking in front of the post office. Everywhere there is dust—it is the summertime—it penetrates one [like] black winter—more penetrating than cold."[24]

It was only after the United States entered World War II and the Historical Section was absorbed by the Office of War Information that southern imagery began to register affirmatively the emergence of a "New South." Southern imagery began to be fashioned that promoted the war effort rather than passed judgment on contemporary society. Neither celebrations of dysfunctional or eccentric premodern cultures nor criticisms of rapid and haphazard modernization were appropriate publicity for a nation at war. The war effort required integration rather than diversity. Pictorial representations of the South needed to fit a template that emphasized achievement. Well before he left the government service for employment with Standard Oil in the fall of 1943, Stryker had begun to represent the United States and its southern region as the "good culture," dynamic, resilient, and offering both affluence and opportunity due to the power of American enterprise and the soundness of government policy.[25]

During the last eighteen months of the Historical Section's existence, the South of this photographic file became born-again American. The region became the site of massive feats of engineering, such as the construction of the "Big Inch" pipeline from Longview, Texas, to Norris City, Illinois, a distance of 550 miles. The twenty-four-inch pipeline was designed to deliver three hundred thousand barrels of crude oil a day and was eventually extended to Philadelphia. In October 1942 John Vachon followed its construction from the Texas state line to Gurdon, Arkansas, as it remorsely cut through cotton and corn fields. The gist of Vachon's series is that the building of the Big Inch was a triumph of coordinated technology and manpower. Vachon carefully detailed the construction process from a machine moving at one mile per hour, which dug a five-foot-deep trench in soft ground in advance of the teams that laid and welded forty-feet sections of pipe, the doping crews who primed the pipe with hot asphalt paint, and the work teams who wrapped it with felt.[26]

The TVA was also represented as a triumph of American mind and muscle. By 1942 Stryker had accumulated substantial holdings on the TVA from the work of the Authority's own photographers. Only when the defense effort made the TVA politically "respectable" was he prepared to assign his own photographers to the TVA region. Rothstein and Delano visited eight dams during the summer of 1942. Unlike the photographers employed by the TVA and the Rural Electrification Administration, who produced, in David Nye's words, an "iconography of the technological sublime," Stryker's photographers were concerned with technical detail in their studies of substations, generators, transformers, transmission yards, and switchyards. Their images are not so much futuristic or majestic as realistic and accessible. They relate to a science of fact rather than of fantasy. This demystification of the technology used to generate electricity had its counterpart in the photographers' renderings of dams under construction. Landscapes of construction sites indicated the tremendous scale of the enterprises, and studies of busy teams of drillers represented the dams as achievements of purposeful human endeavor as well as advanced engineering skill. The South became the site for an epic demonstration of the superior capacities of the nation at war.[27]

To judge from the contents of much of the photographic file for 1942 and 1943, one would think the South had undergone an overnight economic metamorphosis. It had become a hive of industrial activity. In July 1942, at Ingall's Shipbuilding Company at Decatur, Alabama, Delano photographed welders, tackers, pipefitters, and crane operators working day and night shifts to build barges for the Army. The large workforce at Higgins Shipyards in New Orleans, photographed by Vachon in June 1943, was even busier, building cargo vessels; landing craft; and the Hellcat, an experimental patrol torpedo boat. American intervention in World War II allowed the photographers access to the plants of strategic industries. In August 1942, Delano documented the processes of aluminium production at Reynolds Alloys at Sheffield, Alabama, and the various stages in the assembly of a bomber at the Vultee Aircraft Corporation in Nashville, Tennessee. Even the South's most unphotogenic extractive industries could assume daunting proportions when their operations were painstakingly documented and their scale panoramically displayed. The Freeport Sulphur Company at Grand Ecaille, Louisiana, photographed by Vachon in June 1943, was shown to be a great complex of sulphur wells, vats, conveyor belts, and pipelines whose product was carried by barges to the storage vats of Port Sulphur. Indeed, transport was represented as a vital adjunct of the powerful industrial system of which the South had became a part. The trucks in which Vachon rode from their North Carolina depots to points south and west during March 1943 carried not only agricultural produce but aircraft engines; the Greyhound buses on which Esther Bubley rode throughout the upper South in September 1943 carried not only southerners but soldiers and defense workers, some of whom were women, from outside the region. Stryker's photographers depicted a transport system that was erasing the isolation of the South's rural areas and of the South itself.

The significance of this wartime shift in the representational configuration of the South by Stryker's photographic unit is not merely a matter of empirical visual record. Documentary photography, as undertaken by the Historical Section, was never a neutral practice. The social and economic changes captured by Stryker's photographers during wartime had been well underway before 1941, yet had been neglected by the Historical Section. A convergence of interests within the FSA explains why this occurred. The agency itself had a vested interest in representing the South as needing remedial programs and its people as being worthy of public assistance; Stryker's anthropological, documentary interests encouraged the project to focus on premodern survivals in southern society, while the photographers themselves were captivated by those features of southern life that were in contrast to metropolitan, commercial America and stimulated their creative sensibilities. The New Deal at high tide and the burgeoning state that provided its dynamic did not act solely as stifling and restrictive forces on photographic practice. Rather, they generated a fluidity in which political idealism, representational concerns, and bureaucratic requirements clashed as often as they intermeshed and that permitted photographers to accomplish work, although under the auspices of Stryker and in pursuit of the parent agency's objectives, that was not premeditated in the Historical Section's headquarters. The bruised landscape and residual cultures of the South depicted in the Historical Section's images contained an antimony that was typical of cultural representation in the United States during the 1930s and was a product of the project's various influences. They contained both sociology and sensibility and, in equal measure, condemned and celebrated the South's premodernity. As such, the Historical Section's photographs contributed to the enduring myth of a distinctive rural South, intriguing for its contrapuntal themes of misery and innocence.

These tensions in the work of the Historical Section in the South were resolved by American intervention in World War II. By the end of 1941, the Historical Section's representations of the South had begun to emphasize southerners either in harmony with a benign and abundant natural order or taming nature in the cause of military preparedness and national strength. To demonstrate its own usefulness and to ensure its own survival in the context of wartime, the agency and its officers demanded publicity that supported the war effort and required affirmative representations of the South as a component of the arsenal of democracy. Within the Historical Section, issues of patriotic purpose and organizational imperative were reinforced by Stryker's ambitions to make a visual record of the transformations required by a society during wartime. Consequently, the creative latitude previously available to the Historical Section's photographers became more restricted. Only in its later stages did Stryker's unit unequivocally register and affirm an emergent South, and through its concern to provide visual publicity that complemented the war effort, it anticipated the development of a parallel myth of the South that has gained particular currency during the postwar era: that of the erasure particularism and the emergence of the "New South."

NOTES

1. David P. Peeler, *Hope among Us Yet: Social Criticism and Social Solace in Depression America* (Athens: University of Georgia Press, 1987).

2. Jonathan Daniels, "Seeing the South," *Harpers,* November 1941, 600.

3. Numan V. Bartley, "The Era of the New Deal as a Turning Point in Southern History," in *The New Deal and the South,* ed. James C. Cobb and Michael Namorato (Jackson: University of Mississippi Press, 1984), 138–143; Morton Sosna, "More Important Than the Civil War? The Impact of World War II on the South," in *Perspectives on the American South: An Annual Review of Society, Politics and Culture,* vol. 4, ed. James C. Cobb and Charles R. Wilson (New York: Gordon & Breach, 1987), 145–158; Pete Daniel, "Going among Strangers: Southern Reactions to World War II," *Journal of American History* 77 (1990): 886–911.

4. At the end of 1938, with the Historical Section's future secure, Stryker began to reveal his ambitions to transform his unit from a publicity resource

into an archive for American social history and a semiautonomous research agency that resembled a university research institute. See *Washington Post,* December 27, 1938; Stryker to Harry Carman, March 2, 1939, microfilm NDA 8: 406, *Stryker Papers,* Archives of American Art (AAA), Washington, D.C.

5. Interviews of Carl Mydans and Jack and Irene Delano conducted by Richard K. Doud, April 29, 1964, and June 12, 1965, microfilm 716, 555, *Oral History Collections,* AAA; Greg Day, ed., "Folklife and Photography: Bringing the FSA Home. Recollections of the FSA by Jack Delano," *Southern Exposure* 5, nos. 2–3, (1977): 124–125.

6. The classification number of Lee's image is LC-USF 33–11622–M2, and Shahn's is LC-USF 33–2058–M4. In the Southern Series, Shahn's photograph is attributed to Dorothea Lange. However, the date and location make it certain that it belongs to Shahn.

7. "The Story of Escambia Farms," microfilm reel 13: 489, *FSA-OWI Textual Records,* Library of Congress (LC), Washington, D.C.

8. Evans's image is classified as LC-USF 34–28013–A, Rothstein's as LC-USF 34–25017–E, and Post Wolcott's as LC-USF 34–51864–D.

9. Post Wolcott to Stryker, July 5, 28, 29; October 2, 1940, microfilm NDA 30: 181, 333, 398, *Stryker Papers,* AAA.

10. LC-USF 34–51705–D; Arthur F. Raper and Ira De A. Reid, *Sharecroppers All* (Chapel Hill: University of North Carolina Press, 1941), 228.

11. For a fuller discussion of the Historical Section's engagement with agricultural mechanization and modernization, see Stuart Kidd, "The Cultural Politics of Farm Mechanization: Farm Security Administration Photographs of the Southern Landscape, 1935–1943," in *Southern Landscapes,* ed. Tony Badger, Walter Edgar, and Jan Nordby Gretlund (Tübingen, Germany: Stauffenburg Verlag, 1996), 142–154.

12. Pete Daniel, *Breaking the Land: The Transformation of Cotton, Tobacco, and Rice Cultures Since 1880* (Urbana: University of Illinois, 1985), 155–183; Taylor to Thomas C. Blaisdell Jr., June 3, 8, 1937, microfilm reel 15: 503–506, *FSA-OWI Textual Records;* Lange to Stryker, "More Field Notes on Displacement, June 1937;" June 23, 1937, microfilm NDA 30: 809, 821–822, *Stryker Papers,* National Archives, Washington, D.C.

13. The classification number of "Tractored Out" is LC-USF 34–18281–C; Archibald MacLeish, *Land of the Free* (London: Boriswood, 1938), 55.

14. Taylor to Thomas C. Blaisdell Jr., "Note from the Field: Seven Texans of Hardeman County, Texas," June 8, 1937, microfilm reel 15: 503–506, 1646, *FSA-OWI Textual Records.*

15. Harriet L. Herring, "Notes and Suggestions for Photographic Study of the 13 County Sub-Regional Area," Microfilm reel 18: 85, *FSA-OWI Textual Records.*

16. Shahn quoted in John D. Morse, ed., *Ben Shahn* (London: Secker & Warburg, 1972), 136; Post Wolcott to Stryker, "New Orleans," microfilm NDA 30: 301, *Stryker Papers,* AAA.

17. Jonathan Daniels, *A Southerner Discovers the South* (New York: Macmillan, 1938).

18. Wolf, "FSA," January 5, 1940, microfilm NDA 8: 959–960, *Stryker Papers,* LC.

19. Memorandum from Adams to Stryker, n.d.; Notes on San Augustine submitted to *Travel Magazine,* microfilm reel 15: 536, 538, *FSA-OWI Textual Records.*

20. Anderson to Stryker, March 28, 1940, microfilm NDA 8: 759–760, *Stryker Papers,* AAA.

21. Roy Emerson Stryker and Nancy Wood, *In This Proud Land: America 1935–1943 as Seen in the FSA Photographs* (London: Secker & Warburg, 1974), 15.

22. Roy Stryker, "The Lean Thirties," *Harvester World* 51 (February-March 1960): 14; Lange's caption is attached to LC-USF 34–18699–E; Stryker to all photographers, "General Notes for Pictures Needed for the Files," 1939, microfilm NDA 8: 800, *Stryker Papers,* AAA.

23. Delano to Stryker, March 20, May 5, 1941; Collier to Stryker, June 3, 1942, microfilm NDA 25: 1088, 1118; 870, 874, *Stryker Papers,* AAA.

24. George C. Stoney, "Childersburg Still Squats," microfilm reel 14: 97, *FSA-OWI Textual Records;* Collier to Stryker, June 1942, microfilm NDA 25: 872, *Stryker Papers,* AAA.

25. Ulrich Keller, *The Highway as Habitat: A Roy Stryker Documentation, 1943–1955* (Santa Barbara, Calif.: University Art Museum, 1986), 10.

26. Vachon's series is catalogued as *Lot 13,* Reel 2, Prints & Photographs, LC.

27. The TVA images are catalogued in Lots 143–144 and 147–148, Reel 10, and Lots 149–150 and 153–154, Reel 11, Prints & Photographs, LC; "Rothstein Review, vol. 1, no. 1," January 11, 1942, microfilm NDA 26: 53, *Stryker Papers,* AAA; David E. Nye, *Electrifying America: Social Meanings of a New Technology* (Cambridge, Mass.: MIT Press, 1990), 331.

7

Making and Meaning in the English Countryside

CHRISTOPHER BAILEY

When writers in the late nineteenth century wanted to evoke ordinary things, their reference points were generally the artifacts of preindustrialized manufacture. Modernism's rupture with the past was signaled in no sharper way than in the theorist's identification of mass-produced goods, laboratory glassware, or cast iron machine components—rather than willow cricket bats and walking shoes—as the products that typified, or even constituted the essence of, the modern age.

In the various discourses of the interwar period in England, the objects that were made in rural workshops and factories, following vernacular tradition, exhibit an extraordinary lability. Their meaning depends on the speaker and the audience, on whether the process is being privileged above the product, and on the context of both process and product as either picturesque habitat or arena for economic development. The countryside was both of these in the years between the wars, and it is the manner in which rural industry was debated, and its products valued, that gives clues to the emergence of today's landscape of statutorily preserved views and habitats, in which the working countryside is overshadowed by national parks, sites of special scientific interest, and areas of outstanding natural beauty.

A sketch of the parties to the negotiation of meaning could be made as follows: the Rural Industries Bureau (RIB), established by the government to foster development in rural areas through the central application of bureaucratic schemes and policies; the Rural Community Councils, whose interests as community activists intersect with the former in the matter of small-scale manufacturing; and the cultural commentators, whose observations often become the metal of which the hotly debated meanings are forged.

So powerful is the perspective of the cultural commentators that the standard view of the period offers the rural industries organizations only as nostalgic ideologues foisting anachronistic craft activities on unsuspecting country dwellers. Modernity and rurality, it is argued, are defined as antithetical. In arguing that this is a retrospective judgment on the economic prospects of rural areas and that "it needn't have been like that," I show that rural communities have been participants in, rather than victims of, processes of social and economic restructuring. The weighty, but usually discounted, evidence of the effectiveness of rural communities in shaping the concept of rural industry suggests that we must also question the usefulness of the top-down approach of many rural sociologists. I suggest that we replace this orthodoxy with approaches that more readily take account of regional, cultural, and historical difference and examine the interaction of distinct layers of activity, from state to locality. The focus of investigation, to correct the earlier emphasis, ought to be on the interaction of my three "speakers," as "agents in localities." In his agenda for a revised approach to rural sociology, Mormont argues, "Its subject may be defined as the set of processes through which agents construct a vision of the rural suited to their circumstances, define themselves in relation to prevailing social cleavages, and therefore find identity, and through identity, make common cause."[1] Seeing action in context means understanding the accessibility of one agent to another, while networks of interaction can be studied to show how particular places are represented. There is a need to look both "upstream" and "downstream," to explore how networks of agents maintain distinctiveness in relation to others and how more powerful networks condition social change, as well as speaking for the powerless by attempting to impose their representations.

The "rural problem" is usually defined in official documents as a tendency for population to leave the land, even when jobs are available, to the detriment of economic activity in country areas. There have been two principal schools of thought about the most effective remedy to the rural problem. One has regarded the countryside as a garden under the stewardship of landowner and farmer, with agriculture the principal economic activity. According to the other view, town and country face similar development needs, which might be met by the introduction of new forms of employment. The roots of both attitudes can be traced to the nineteenth century, but by the 1940s the ascendancy of those who saw agriculture as the only appropriate activity for rural areas was assured.

The formal expression of this position is found in a royal commission, set up in the light of new attempts comprehensively to map land use in the British Isles, out of which emerged the dominant vision of rural England.[2] Between 1941 and 1942 the Committee on Land Utilisation in Rural Areas, chaired by Lord Justice Scott, took evidence from many bodies with an interest in development and reconstruction, including the RIB.[3] The report was written in the context of a rigorous command-driven agricultural economy, the wartime "Dig for Victory" campaign. Looking forward to peace, the commission proposed a single, shared vision of the modern countryside. However, anyone expecting the committee to look other than to farming for the "well-being of rural communities" found his or her hopes dashed by the final report. Scott, who had been vice president of the Council for the Preservation of Rural England, wrote that farms should be regarded as the "nation's landscape gardens." The value of agriculture as the major economic activity was that there was in it "no antagonism between use and beauty."[4] Any nonagricultural use should be an "appropriate" one, which in effect meant the extractive industries, such as mining and quarrying, and "rural trades and crafts." Many of the report's recommendations concerned the aesthetics of landscape, and it is easy to see in embryo in Scott the draconian controls over industrial development ushered in by the 1947 Town and Country Planning Act. Under a comprehensive national land use policy, rural areas had become, so to speak, "zoned" for agriculture and recreation (Figure 7.1).

The majority view of the Committee on Land Utilisation in Rural Areas did not go unchallenged. The economist S. R. Dennison disputed the logic of the final report issuing his own minority report. In it, he condenses Scott's argument about the motivation of rural emigrants for deserting the land:

> The only meaning which I can give to [the Report] is that rural workers must first be given as high standards as the industrial workers, and indus-

Figure 7.1 Cutting hoops. No date. Published in *The Times* (London). Courtesy of RIB.

> try can then be introduced; otherwise the rural population will become discontented at the sight of higher standards than they themselves enjoy, and will move away from agriculture. And the logical inference is that if they cannot be given such high standards, then industry must be kept away. The Majority, indeed, expects that agriculture can give the necessary standards; but this throws us back once more on our fundamental disagreement on the question of the prosperity of a large agriculture.

Dennison took issue with the wisdom of enlarging the agricultural sector at the expense of the manufacturing and service sectors and made a concerted plea for measured industrial expansion in rural areas. If rural areas were to be thus "protected," Dennison believed, they would lose out twice over. "Here the Majority gets the worst of both worlds; industry stays out of the villages so as to avoid disturbing them, but the workers might still go to work in the nearby towns."[5] The definition of the "rural problem" as one of preservation, as much aesthetic as social, received powerful support from the newly established profession of town planning. Like the private lobby groups, such as the Council for the Preservation of Rural England, the planning fraternity sought to insert itself between the forces of industry and the precious form of the countryside, offering "rational" aesthetic approaches to what had hitherto been left to the working of Victorian market forces. In his description of the landscape of northeastern England, *"Hills and Hells,"* the planner and theorist Thomas Sharp is struck by the contrast evident in the beauty of the "fine machines" turned out in the "mixture of drab buildings . . . spread thick over a wide terrain till every vestige of physical beauty and seemliness had disappeared from the scene."[6]

For the present argument, perhaps the most telling of Dennison's points is his reminder that "our duty is to foster the well-being of rural communities, and not to 'preserve' them."[7] As we shall see, this objective matches closely the founding intentions of the Rural Community Council movement. The ambivalence of the RIB, however, was such that when it reviewed the report in its magazine *Rural Industries,* it endorsed not Dennison's but the majority's position.

Compared with organizations dealing with "urban" manufacture, such as the Design & Industries Association and its successors, the RIB and its associated county-based organizations have received little attention from historians.[8] Yet there are strong thematic links between the bureau's magazine *Rural Industries* and the publications of the Design & Industries Association, the Council for the Preservation of Rural England, and thence the vast popular literature about the countryside. Comparison of this literature reveals that the RIB was equivocal, but admitted the potential for being modern in the countryside in ways that none of the other organizations did. These obscured perceptions of the countryside were represented in the work of the rural industries organizations, including their working documents, publicity, exhibitions, and publications.

The loose partnership of bureaucracies, often dubbed the "rural industries organizations," which emerged between the wars, consisted of the RIB; the Rural Community Councils; and the offshoots of the councils, the Rural Industries Committees, which were partnerships between the Rural Community Councils and the County Councils.[9] The RIB was funded by a grant from the government's Development Commission, itself created as a result of the Road Fund and Development Act of 1909, to "promote the economic development of the UK and the improvement of roads therein." The decade after the end of World War I also saw the formation of a number of organizations such as the National Farmers' Union and the Women's Institutes in response to the destabilization and loss of social cohesion widely experienced in rural Britain. There was thus a formidable range of opportunities for engagement in social action in the countryside.

As the major source of state advice and information to businesses in rural areas, the RIB had a determining influence on policy. Yet commentators have often pointed to conflicts or lack of coherence in the bureau's objectives and actions. The RIB's declared aim was not to propagandize to the public but to be engaged in successful economic intervention in rural communities. The bureau even criticized the Council for the Preservation of Rural England for its faith in "the old, forlorn process of educating the public."[10] In a recent interview, Marianne Straub, whose commissioned report on Welsh textiles led to a revival of the fortunes of a number of mills in the 1930s, lent weight to the view that unlike the design reformers of the Council for the Preservation of Rural England and the Design & Industries Association, the RIB was not primarily concerned with propaganda. "The RIB was a much more active thing. The DIA was a sort of association, much more theoretical . . . I must stress that what the Bureau was helping here was *industry* not craft . . . It was not interested in people like Ethel Mairet and Bernard Leach, it was interested in *rural industry.*"[11]

This raises the question of what exactly the RIB was to do to achieve its aims. Much depended on the definition of rural industry. In 1918, J. L. Etty's report to the Development Commission on the subject had drawn attention to the problem of definition. Did it mean traditional village handicrafts, many of which were dead or dying, and if it were to encompass more, should it include the builder, dressmaker, and baker, or the larger-scale industries that happened to be located in rural areas?[12] By 1922 the bureau, clearly mindful of what its limited funding would permit, had formulated its own definition: "industries and handicrafts carried on in rural districts either in the home or in small workshops and factories."[13] In light of the bureau's grant from the Development Commission and the scale of the economic problems it faced, this remit seems no more than realistic. Nonetheless, the RIB was clearly open to the charge, laid by W. M. Williams's influential 1958 account, that the typical bureau industry would be "small in size, have a strong element of craftsmanship, and a definite geographical location."[14] Williams attributes the bureau's impotence to its unscientific and romanticized definition of industry, its failure to encourage economic strategies such as relocation of industry to rural areas, and its neglect of proper surveys or research. Much of the criticism of lack of economic influence is justified. The RIB was not empowered to trade, so its intervention was inevitably at the level of exhortation. This weakness was not remedied until the establishment of the Rural Industries Loan Fund in 1940.

The RIB did indeed focus largely on the crafts, as recent research shows, and what could be done to "improve" them by standardizing the design of handmade products. The bureau's budget was structured around a list of crafts such as ironwork, woodwork, basketry, textiles, and quilting. As a list of fourteen Bureau pamphlets brought out by 1924 shows, however, the range of economically viable activities envisaged in rural areas was altogether more inclusive. In addition to encouraging smiths to look for new uses for their skills, pamphlets suggested making "straw-rope" (a useful substitute for the temporarily unavailable sisal), breeding angora rabbits for fur, and making peat firelighters.[15]

In its first three years, the bureau, assisted by additional Development Commission funds, tackled the question of training smiths, wheelwrights, and carpenters in mechanical repair work to replace the rapidly declining demand for work on horse-drawn equipment. Oxy-acetylene welding, emery grinding, and drilling machines were demonstrated to groups of smiths in Oxfordshire, Sussex, Winchester, and Leeds, and model workshop layouts were published in *Rural Industries.* Vans were bought to take equipment to smiths who had none, and competent workers were included in a list made available to the Royal Institute of British Architects, so that they might make use of traditional materials and skills in new buildings (Figure 7.2).

Alongside the training and re-equipping of craftworkers, the Bureau promoted higher standards of design through pattern books and drawings made by professional designers and worked to increase sales through the application of marketing techniques for crafts products. Most visibly, it undertook a campaign of persuasion, through its reports on "revived" industries, its advice pamphlets, and *Rural Industries.*[16]

From the RIB's London office, the director managed the work of around a half dozen specialist advisers, each of whom provided advice on an aspect of rural crafts and trades. The raising of standards was taken very seriously by the RIB, which secured funding for two "design research studentships" at the Royal College of Art in London. In follow-

Figure 7.2 A modern workshop, equipped and laid out to RIB design and showing both maintenance work and sidelines manufacture being undertaken. No date. Photographer unknown. Courtesy of RIB.

ing years, a bank of these and other designs was built up, together with a library, for the instruction of rural craftworkers. The debate about standards demonstrates clearly the extent to which the bureau shared the opinions and tastes of the design reform movement and, more crucially for my argument, the ways in which design aesthetics functioned as a weapon in the struggle for influence between the London-based RIB and the widely dispersed Rural Community Councils and their allies. The idea that design is an economic panacea has a long pedigree. In putting it forward to support the "sidelines" policy aimed at rural smiths, the RIB was simply persisting in a belief shared by many others. George Marston's articles on design, published in the 1920s, and those by bureau advisers, such as W. A. Elwood in 1933, attempted to improve standards by example, introducing a vocabulary of criticism derived from art education. The guiding principle promoted in these essays was the Ruskinian one, that "design is structural invention rather than applied decoration."[17]

Friction between centrally determined standards and local practice occurred when the bureau, through including work on its exhibition stands around the country, appeared to give its endorsement to the work its advice and support had fostered. *Rural Industries* frequently gave room to contributors to complain about the standard of work at county shows and industrial exhibitions. Of the stand at the 1927 British Industries Fair, one writer was "frankly of the opinion that too great a proportion is bad." The county shows often came in for criticism, either in commissioned articles or in the correspondence columns. Having in its earliest statements committed itself to improving design, the RIB found itself involved in a debate about relative values. As the bureau admits in its Fifth Annual Report, its attempts to persuade blacksmiths (there were about 180 across the country on the register) to adopt approved designs met with a mixed reception. Many smiths resented "museum designs" being imposed on them, often without reference to commercial factors. Others simply disliked what they regarded as "art stuff."[18] The bureau rather defensively tried to explain itself to its subscribers. It had never, it said,

> imposed any standard of design or taste upon craftsmen, who it has aided, beyond insisting on the recognition of one or two elementary principles:
>
> 1. that the craftsman should make things for use not ornament;
>
> 2. that the construction should be controlled by the use to which the article is put, and that the final form must be dictated by the character of the material.[19]

These denials are not consistent with the evidence presented two years later by the RIB to the Rural Industries Enquiry in 1930. Its director, J. R. Brooke, voiced his concern about the role of the county organizers for rural industries. He argued that the organizer, who had necessarily worked closely with a craftworker, could not then be expected to come to an objective conclusion about the standard of the work produced. The organizers were generally employed by the County Councils in conjunction with the Rural Community Councils. Brooke, who had lately seen firsthand how the Swedes had developed a national network of handicraft societies, believed that their designs were successful because of a centralized aesthetic policy. The Rural Community Councils, he argued, had learned a little from the bureau, but their lack of technical and aesthetic expertise had "resulted in little progress, and in some cases in actual deterioration," and he concluded that "[i]t is now essential that the Bureau be responsible for Rural Industries Policy and standards." Marston agreed that a firm hand was required, stating, "If we are to make progress, it is necessary for us to be able to impose our views as to what is good design and technique."[20] This could best be done, argued Marston, if greater numbers of technical advisers were employed centrally to take the bureau's policy out to the counties.

The standards debate masks a tension. It is a dual tension between a London bureaucracy and a highly complex and diverse regional network, and between an absolute aesthetic standard and a relativistic system of values built on local knowledge. The RIB's overt objective was the extension of its aesthetic influence through an increased budget. At the same time, we can see here the resolution of a new image of the crafts in which a higher value was placed on technical accomplishment than on usefulness, and continuity of artistic tradition was prized above economic viability. The diverse economies still identified as a possibility by Dennison in 1942 were being constrained to what was deemed "appropriate" activity in the terms of the Scott Report.

The readers of *Rural Industries* and the organizers who encouraged them in their work occupied a social position quite unlike that of the interested middle-class disciples of design reform. They were producers as well as consumers of the rural perspective. As interest and investment in the countryside by the professions, and by an increasing number of middle-class residents, converted the countryside into a preservation area protected from progress, there was a noticeable resistance that asserted the validity of local knowledge over imposed values. The compromises that appear in print in *Rural Industries* were often an acknowledgment that, to the rural industries organizer out in the field, things might look very different. As a Rural Community Council chairman bluntly informed the Rural Industries Enquiry in 1930, he was against the imposition of bureau "standards" in craft. "As for keeping people up to the mark, we have not insisted on too high a standard because the men have suffered badly and must not be too hardly used . . . we want to get the men's confidence."[21] It would be unwise to press too far the case that the Rural Community Councils represent a consistently radical strain of thought about the countryside. Generally they saw their role as the support of the individual craftworker in the context of the well-being of the community. In doing so, the councils might well emphasize marketing rather than standards, women's "leisure" crafts over "men's work," and local autonomy over central policy. Just as the RIB's position resists easy categorization, the Rural Community Councils were diverse in constitution and their members acted from a range of motives. Between 1922 and 1942, a total of twenty four councils, covering twenty nine counties in England and Wales, were created. Their remit widened as they took over many of the functions of the parish councils that in so many areas were moribund by this date.

Depending on the expertise and views of its members, a Rural Community Council's activities might include an enquiry into the adequacy of village halls and playing fields, tours of the village concert party, assistance in providing lantern lectures, consideration of questions affecting juvenile life, and the formation of village councils. The stimulation of rural industry was only one weapon in the armory and was unlikely to be deployed solely on narrow economic grounds. The speech by Sir Henry Rew at the first Rural Life Conference in Oxford in April 1920 put this point across to an audience of the already converted. "The time has come when the human needs of the countryside have become insistent, and the future of agriculture is seen to involve a sociological, as well as economic, problem."[22] Where previously the local "squirearchy" might have supplied the deficit, the impact of the social changes I have already described and the loss of the "natural" leaders during World War I led to a vacuum, filled by a range of largely middle-class activists. The men and women who sat on the Rural Community Councils were, in the case of many counties, fairly described by James Noel White as "landed and wealthy, . . . able to assist financially [and] in a way which remains unrecorded, personal and private."[23] The first Rural Community Council, founded in Oxfordshire in 1920, brought together representatives of the Oxfordshire and the National Federation of Women's Institutes, the National Council of Social Service (NCSS), the Free Church, University Tutorial Classes, and the Worker's Educational Association.[24] A network of like-minded people spread Rural Community Councils across the counties of southern England and the Midlands. The education network, especially adult education, provided a useful pool of able and energetic evangelists for the cause. This range of interests in recreation, adult self-improvement, and education was typical, not only of the councils, but also of the participants at the series of rural life conferences that was promoted by the NCSS during the interwar years.[25] Participants in these discussions felt that it was natural to talk about village life in terms of "cultural deprivation" and that it fell to them to "give the initial prescription and supervise the treatment." Management of rural communities by the professions thus brought in its wake the new terminology of "latency" and "development," borrowed from the social services (Figure 7.3).[26]

In most cases, the Rural Community Council set up a Rural Industries Committee, often with representation from the County Councils and from locally respected craftworkers. In Gloucestershire, the Rural

FIGURE 7.3 The 1928–1929 annual report for the Rural Community Council of County Durham, United Kingdom.

Community Council included craftworkers such as George Hart, who was described as coming "from an interesting colony established in the Cotswolds where rural handicrafts were practised."[27] By 1928 demands for support were such that the Rural Industries Bureau thought it was time to take a hand in determining the work of the Rural Community Councils. In its *Memorandum on Rural Industries,* the RIB argued that the Rural Community Councils should take responsibility for rural industries work. It defined this narrowly as the trades ancillary to agriculture, country crafts, dependent on "local conditions or supplies of materials," and "occupations which may help to increase the standard of home life" or, in other words, the "domestic crafts."[28]

As the Rural Industries Committees proliferated, the RIB became more reliant on the county organizers for reports of activity and, as we have seen, often found itself in conflict with the approach and values applied at the county level. Whenever they were confronted with demands for conformity, the county organizers insisted that there was no blueprint for a Rural Community Council and successfully resisted attempts by the Development Commission to grade their performance by a common standard.

Just as contact with rural craftworkers in the 1950s convinced W. M. Williams that few of them were concerned with abstract debate, as the RIB might have wished, so there was a mood of realism in the Rural Community Council–sponsored group discussions held around the country. As Williams soberly observed, rural craftworkers, like their urban counterparts, were "more interested in collecting outstanding accounts than in the poetry of Nature."[29]

The potential for open-mindedness about economic strategies for traditional crafts is well demonstrated by a vivid account of the philosophy of the Rural Community Council in Cambridge by its secretary, E. R. Vincent. His prescription for the rural problem is an unromantic cure or kill. "If the village blacksmith will not adopt modern methods, will not make the best use of modern machinery, will not meet a changing market, he will disappear, and no amount of sitting under chestnut trees and singing in the choir will save him. And we should not want to save him."[30] Through modern marketing techniques the business prospects of the good smith might be transformed. A Rural Community Council, Vincent argued, "must adopt the methods of the psychoanalyst rather than the surgeon," trying to understand the mind of the consumer, rather than crudely amputating living craft skills. Why, he asked, should not the blacksmith create a demand, say, by setting up a collectors' club for horseshoes, to be used as amulets or table decorations? Vincent's economic model, the recent success of chewing gum manufacturers in building a new market from an apparently useless vegetable material, strikes a discordant note with the RIB's high-minded insistence on the production of "useful" goods, fit for their purpose. Under the circumstances, he argued, marketing cooperatives and imaginative selling techniques were a more appropriate response.

In the 1920s, as craftworkers' basic trade declined, they had first been advised to diversify. Through this strategy craftworkers could scrape through hard times by making toys or ornamental scrollwork. This must account for the enormous variety of work that was exhibited at county shows and caused the RIB so much anguish about standards. By 1936 the Gloucestershire County Annual Report shows this policy declining in favor of support directed to industries for which there was a predictable market.[31] Thus basket making was supported because there was (and remains) no mechanized competition. Instead, the focus was on an improvement in marketing methods, easily supplied through a federation of makers and an agent. The charcoal burners of the Wyre Forest were given access to an existing market: the many leisure artists who were forced to use imported French charcoal. The regular rounds of advisory visits by rural industries organizers and technical advisers served to provide the networking that Vincent had graphically described in 1927.[32]

Under its first secretary, Madox Yorke, the Gloucestershire Rural Community Council helped Edward Gardiner revive the making of Gimson chairs. After Gardiner's death, Neville Neal took over the workshop, and he found that Yorke's successor, T. G. Castle, could cover expenses for him to attend several shows, including the county and royal shows and Olympia, each year to promote his work. Distribution through well-known retailers could also work well. According to Neal, Barrow's Department Store in Birmingham "took everything Gardiner made before the War."[33] This connection was a natural one, given that Barrow's displays of rural crafts were established and organized by Yorke after his resignation from the Gloucestershire Rural Community Council.

The range of work shown at Barrow's and other stores was considerable, covering ceramics, glass, textiles, and furniture. Here, too, the RIB found that work chosen for display failed to meet their standard. A bureau report repeats the overheard remark of a visitor at a ceramics exhibition: "In the Potteries that'd be a second."[34] It is not so clear what the bureau felt about the mass-produced goods on display in other departments, but at the end of the war, Barrow's discontinued its policy of exhibiting local craft products. When it did show crafts, the store favored displays of Scandinavian hand-produced goods.

There was an understandable anxiety, which is not new in the 1920s, that the crafts, practiced for enjoyment or for pin money, would make it harder rather than easier to improve the economic viability of rural industries. In the interwar period, as Pat Kirkham has shown, the spread of ready-made patterns, cheaper materials, and a network of women's organizations such as the Townswomen's Guilds and the Women's Institutes, made the rural market highly attractive to companies like Dryad, which produced the tools and raw materials for the leisure craftworker.[35] *Rural Industries,* like many mass-market magazines, carried ads for Dryad's products, although in its columns there was often criticism of the craft goods which emerged as a result. In 1926, Marston, then a technical advisor to the RIB, wrote that the craft revival was taking place not among the artisan class, but among the leisured class. He argued that unless "marketing is got right," leisure crafts products, by definition largely made by women, will "damage the market."[36]

The value of the so-called leisure crafts in feeding or clothing rural communities appears not to have been considered as seriously by the RIB as the contribution to the cash economy made by "professional," that is, predominantly male, craftwork. Despite the representation of the National Federation of Women's Institutes on the RIB's Central Committee, the inclusion of the county federation on every Rural Community Council, and the existence from 1922 of a high-powered Women's Advisory Committee at the bureau, there was no challenge to the assumption that women's craft activities were essentially a leisure occupation and therefore a threat to what the bureau regarded as their real object. When craftwork was shown under the RIB's banner, this gender distinction was likely to be reinforced. Kirkham's conclusion that the Women's Institutes "never challenged the dominant ideology relating to women's role in society" applies equally to the RIB. While the Women's Institutes expanded, through their classes and exhibitions, what women might be allowed to learn, their products continued to be shown on stands describing women as amateurs pursuing their craft as a recreation. In other circumstances, such as in areas where the principal male occupation was neither in agriculture nor in ancillary trades and where there was very high male unemployment, the bureau showed direct support for women's domestic crafts. In the Welsh valleys and the northeast of England, the female crafts of textiles and quilting accounted for some of the bureau's greatest successes.[37] National publicity for the quilting revival resulted in examples being purchased by the Victoria and Albert Museum and in sellout displays of work in the RIB's London retail outlets.

In a paper given in Cambridge in 1982, the anthropologist Anthony Cohen argued that the struggle for economic and social viability being waged by rural communities should not be dismissed as anachronistic parochialism. From an ethnographic perspective, local knowledge represented an "indigenous assertiveness," in opposition to the "celebration of rusticity by outsiders."[38] As we have seen, the promise of progress was steadily drained from rural communities in the interwar period, leaving a space in which the preservationists could inscribe their own meanings. Whatever the successes and failures, in economic or social terms, of the Rural Industries Committees, the countryside was undergoing a shift in its representation such that the RIB found itself swept into the same camp as the preservationists. The changing definition of rural handwork, offered and amended by the bureau, contested and varied endlessly by a legion of men and women in rural areas, was eventually pressed into the mold of the public imagination as a rustic survival.

In the wave of popular writing about the countryside, including tour guides and memoirs such as the immensely successful *Lark Rise to Candleford* trilogy by Flora Thompson, and the reports of the Council for the Preservation of Rural England and the Design & Industries Association, there are some common themes. Change in the countryside is linked with loss. In many cases it is an unpeopled landscape that is described. Often, in the absence of people, objects are used to symbolize the processes of change. The Design & Industries Association's yearbook, *The Face of the Land,* presents the downfall of the environment in apocalyptic terms. Although the book deals with both town and country, the sense of loss is acutely focused on rural imagery. "And to what have we lost it, this comely, ordered arcady with its gracious market towns and homely villages, as instinct with local individuality and tradition as the speech of their inhabitants? For what have we exchanged these good, these very English things?"[39]

This echoes strongly the sentiment of the Council for the Preservation of Rural England, founded four years before. References to the beauty of the form of the landscape are strongly gendered. The countryside is variously described as "comely," "gracious," or "slatternly" (where "slovenly" would describe a male character), while the march of the town is a "violation." "It is thus that rural England is being dissolved away, disappearing bit by bit, but with a frightening acceleration behind a screen of roadside villas, tea shacks, petrol depots and billboards that violate every canon of seemliness and order and make a slattern of a countryside once serenely gracious."[40] The sense of loss, the picturesque character of the countryside, and the gendering of the imagery used to describe the process of change all conspire to create in the reader's mind a view of the countryside as passive victim, rendered helpless by the progress of the town.

In this lapsarian modern world, change in the countryside can only be for the worse. Indeed, the very word "change" becomes a token of displaced longing in this literature, from George Bourne's *Change in the Village,* to the final chapter of Thompson's *Lark Rise,* also called "Change in the Village," to Thomas Hennell's *Change on the Farm.* An elegiac passage from Hennell's book invests objects with a grand theme of decline.

> All the country's history, and not only a chronicle of small beer, is written out in the carpentry of broken carts and wagons, on the knots and joints of old orchard trees, among the tattered ribs of decaying barns, and in the buried ancestral furrows and courses which can still be traced under the turf when the sun falls slantwise across the fields in the long autumn afternoon.[41]

The replacement of a peopled landscape with a museum of "bygones" was literally achieved in an expensively printed book from 1939 written by H. J. Massingham and illustrated by Hennell.[42] In *Country Relics,* Massingham set out to construct a narrative for his "inventory" of around 120 artifacts relating to rural trades and occupations that he kept in a virtual shrine to past days known as the Hermitage (after the example of the naturalist Gilbert White). The artifacts are grouped in the book into sixteen chapters titled "Stonework," "Timber," "Straw and Basket-work," "The Fireplace," and so on. Around them the author weaves anecdotal material from heterodox sources including sociological, ethnographic, and economic studies. In a number of sketches, he

Timber

the headlands of the open fields. Seedlings had grown into trees with rambling trunks; goosegrass, Jack-run-the-hedge,

The old hedger, grandson of Samson and Fairlinda Dixon. See p. 209

ground-ivy and field grasses yearly rose from the graves of their dead selves, and briars snaked through the tangle in shoots so wanton that only the hedgehog could know their

46

FIGURE 7.4 The bodger of the beech woods, 1939. Illustration by T. Hennell. From H. J. Massingham, *Country Relics,* 1939.

explores the fates of individual craftworkers, such as "the bodger of the beech woods," drawing on K. S. Wood's 1921 study for the Agricultural Economics Research Institute. To this he links a discussion of what hand tools can achieve and compares hand- and machine-turned chair legs to demonstrate the moral and aesthetic superiority of the former. For him, the restoration of wholeness in rural communities was linked with the prosperity of one-man businesses and the principle of inheritance from father to son. The correct approach to production requires control over all stages from the production of raw materials to finished product. Above all, Massingham argued that no craft, including agriculture, could be mechanized. At root, Massingham's appeal was based not on economic but on ideological grounds (Figure 7.4).

Economic factors were, then, only a partial influence on the debate about the place of industry in rural areas. The debate was also shaped, I have argued, by a range of statutory and voluntary bodies concerned with social action in the countryside. These groups, in turn, drew on a range of lived experiences that they fitted to the available forms of description, whether these were the language of reports, literary or anecdotal description, or even photographic imagery. There was, as I have shown, an ambiguity in these positions that permitted a degree of autonomy in defining rural work. Within the debate there was initially an affinity between the experience of the countryside and ideas of progress, but this was steadily replaced, for the professions, by the imposed orderliness of planning, and for the wider public, by the sentimental landscape of nostalgia. Neither vision acknowledged the constructive social or economic potential for craft-based industry.

NOTES

1. M. Mormont, "Who Is Rural? or How to Be Rural," in *Rural Re-Structuring: Global Processes and Their Responses,* ed. T. Marsden, P. Lowe, and S. Whatmore (London: David Fulton, 1990), 41.

2. S. Rycroft and D. Cosgrove, "Mapping the Modern Nation: Dudley Stamp and the Land Utilisation Survey," *History Workshop Journal* 40 (Fall 1995): 91–105.

3. Also listed as giving evidence was the West Midlands Group for Post-War Reconstruction and Recovery, whose committee included P. Sargant Florence, supervisor of the research that led to Nikolaus Pevsner's *Enquiry into Industrial Art in England,* published in Cambridge 1937.

4. Lord Justice Scott, *Report of the Commission on Land Utilisation in Rural Areas* (London: HMSO, 1942), 94.

5. S. R. Dennison, *Commission on Land Utilisation in Rural Areas,* Minority Report (London: HMSO, 1942), 111, 113.

6. T. Sharp, "The North-east—Hills and Hells," in *Britain and the Beast,* ed. C. Williams-Ellis (London: Reader's Union with J. M. Dent, 1938), 143.

7. Dennison, *Commission on Land Utilisation,* 113.

8. The RIB was founded in 1921 with a grant from the government's Development Fund "for the purpose of assisting the development of rural industries, by giving information and advice on matters connected with them." RIB, *Its Objects and Work* (London: RIB, n.d.), 4. The Rural Community Councils were started on the initiative of the National Council of Social Service, itself founded in 1919, to assist in the development of rural life. RIB, *A Memorandum on Rural Industries, Containing a Suggested Programme of Work for Rural Community Councils* (London: RIB, 1928), 1–13.

9. There is no full-length historical treatment of the RIB. A summary of the work of the bureau and other rural industries organizations can be found in W. M. Williams, *The Country Craftsman: A Study of Some Rural Crafts and the Rural Industries Organisations in England,* Dartington Hall Studies in Rural Sociology, vol. 2 (London: Routledge, Kegan Paul, 1958), and in H. Tebbutt, "Industry and Anti-Industry—the Rural Industries Bureau—Its Objects and Work" (Master's thesis, Victoria & Albert Museum/Royal College of Art, London, 1990). Other useful accounts are S. Clarkson, "Jobs in the Countryside: Some Aspects of the Work of the Rural Industries Bureau and the Council for Small Industries in Rural Areas, 1910–1979" (Occasional Paper, Wye College, Ashford, Kent, England, 1980); A. R. Pittwood, "Rural Industries in England and Wales Since 1900" (manuscript, Special Subject Part III, Bachelor of Science in Agricultural Economics, Reading University, 1972). For the history of the National Council for Social Service from its formation, see M. Brasnett, *Voluntary Social Action, a History of the National Council of Social Service* (London: Bedford Square Press, 1969).

10. "The Rural Boom," *Rural Industries* 6 (Spring 1927): 1.

11. From an interview with Holly Tebbutt, in Tebbutt, "Industry and Anti-Industry," 135. Despite Straub's assertion, which is corroborated, for example, by RIB support for the mechanized textiles industry in Wales, the magazine published pictures and ran stories about Leach, Mairet, and Cardew, the last shown in a piece titled "Peasant Pottery" by W. F. Crittall, *Rural Industries* (Summer 1936): 24–27.

12. J. L. Etty, *Report on Village and Rural Industries: Their Condition and Suggestions for Their Organisation* (London: HMSO, 1918), quoted in Tebbutt, "Industry and Anti-Industry," 23.

13. Rural Development Commission, *Report of the Development Commissioners* (London: HMSO, 1922), quoted in Tebbutt, "Industry and Anti-Industry," 23.

14. W. M. Williams, *The Country Craftsman: A Study of Some Rural Crafts and the Rural Industries Organisation in England,* Dartington Hall Studies in Rural Sociology, vol. 2 (Routledge: Kegan Paul, 1958), 8.

15. The full range of publications is listed in the *First Report of the Rural Industries Bureau,* contained within the *Development Commission Report,* no. 14, 1924 (London: HMSO, 1924). About thirty thousand pamphlets were distributed.

16. *Rural Industries* was published quarterly initially at the price of 1s, between 1925 and 1939.

17. "Design," *Rural Industries* (Spring 1935): 4.

18. *Fifth Annual Report of the Rural Industries Bureau* (London: Rural Industries Bureau, 1928).

19. Quoted in Tebbutt, "Industry and Anti-Industry," 101.

20. Quoted in Tebbutt, "Industry and Anti-Industry," 91–94, an account that draws usefully on "Rural Industries Enquiry," Public Record Office, KEW Catalogue Reference D4/421 xc 157810.

21. Evidence given by Colonel Swayne, president of Hertfordshire Rural Community Council, quoted in Tebbutt, "Industry and Anti-Industry," 92.

22. Action for Communities in Rural England (ACRE), Box: "Rural Life, Sir Henry Rew, Chairman of the Village Clubs' Association," typescript in a collection, *Notes on the Conference, Reconstruction and Rural Life* (Cirencester, England: ACRE Archive, 1922). The conference, which was concerned mostly with drafting the constitution for the Oxfordshire Rural Community Council, also produced a "Memorandum on Rural Development" which stressed "the need for a clear association of official and voluntary bodies at local and national level." See "Memorandum on Rural Development, with Special Reference to County Organisation," published by the National Council for Social Service (London, May 1922).

23. Quoted in Tebbutt, "Industry and Anti-Industry," 114.

24. ACRE, Box: "RCC History, Formation of Rural Community Councils," typed notes presumed to be by W. G. S. Adams, c. 1948. In Essex, A. H. Mackmurdo, the founder of the Art Workers' Guild and described by Adams as a "country lover and architect," set out to secure a grant for his village hall and found himself with backing from the Lord Lieutenant Brigadier General Sir Richard Colvin and W. J. Courtaull for the formation of a Rural Community Council.

25. The formalization of crafts training outside the main urban centers can be traced to the establishment of the Cottage Arts Association in 1884. In the following year, inspired by the attitudes toward "amateur work" that G. S. Leland had seen in North America, the renamed Home Arts and Industries Association (HAIA) pledged to "revive cottage industries through [the] organisation of classes for workers in rural districts." The HAIA promoted a "South Kensington" model of instruction, using handbooks of designs, but also directly advised on techniques and tools. By the 1890s, the impact of the association,

especially for the rural working class, was considerable, including countrywide craft classes; an annual exhibition at Albert Hall a sales outlet in New Bond Street, London; and a marketing network jointly operated with the Rural Industries Co-operative Society. Two independent Schools of Industrial Art, at Keswick in the Lake District and at Southwold on the Suffolk coast, both areas with considerable tourist appeal, were linked to the HAIA. The latter school operated a savings bank and a library, thus paralleling the community development focus of C. R. Ashbee's Guild and School of Handicraft. G. Millar, "Controversy over Design Education in England, c. 1864–1911" vol. 2 (master's thesis, University of East Anglia, Norwich, England 1975), 126–150.

26. D. Matless, "Ordering the Land: The 'Preservation' of the English Countryside 1918–39" (Ph.D. diss., University of Nottingham, Nottingham, England, 1989), 6–7. Matless draws attention to the operation of the relative activity and the level of professional expertise, whether technical or aesthetic, among Rural Community Council committee members compared with the perceived latency of the village resident, to produce an apparent neutrality in the policies and attitudes adopted toward the development of community and education.

27. Adams, "RCC History, Formation." The "colony" is presumably a reference to Ashbee's Guild in Chipping Campden. George Hart, the silversmith and brother of Will Hart adopted the registered mark "G of H" for his silverware after 1912 and continued in business in the village.

28. RIB, *A Memorandum on Rural Industries Containing a Suggested Programme of Work for Rural Community Councils* (London: RIB, 1928), 6.

29. W. M. Williams, *The Country Craftsman,* 93.

30. ACRE, Box: "RCC History," notes of a Conference of Rural Community Councils held at Cambridge in 1927.

31. ACRE, Box: *Gloucestershire Rural Community Council Annual Report for 1936.*

32. In his talk, Vincent describes a meeting of fifty smiths and other tradesmen in Cambridge (ACRE, Box: "RCC History"). His claim that it was "the first meeting of rural craftsmen in Cambridge since the Middle Ages" may be an overstatement, but does indicate the importance of Rural Community Councils in overcoming the profound problems of isolation in order to enable craftworkers to become competitive producers. An indication of the level of activity of rural industries organizers is given in the *Gloucestershire Rural Community Council Annual Report for 1930/31* (Gloucestershire: Gloucestershire RCC), 11–12. D. W. Lee Browne, who worked one day each week as an organizer, made three hundred visits to craftworkers, concentrating on smiths in the Forest of Dean.

33. Neville Neal, in conversation with the author, Stockton, Northants, on 2 July 1994.

34. A Birmingham Showroom for Rural Industries, *Rural Industries* (Winter 1931), 7.

35. P. Kirkham, "Women and the Inter-war Handicrafts Revival," in *A View from the Interior: Feminism, Women and Design,* ed. J. Attfield and P. Kirkham (London: Women's Press, 1989).

36. G. E. Marston, "The Women's Institutes and the Handicraft Movement," *Rural Industries* 2 (Spring 1926), 5–6.

37. The Women's Advisory Committee of the RIB undertook a seven-day tour of South Wales, after which it concluded that the RIB should sponsor the "revival and development" of quilting. The subsequent exhibition in London in October 1927 produced sales worth £500. Classes were set up in the "neediest mining districts" of Monmouth and Glamorgan from which were produced quilts that were shown regularly in the showroom of County Industries Limited at 26 Ecclestone Street, London. See RIB, *Fifth Annual Report of the Rural Industries Bureau* (London: RIB, 1928).

38. A. P. Cohen, "Anthropology and the Culture of Localism in Britain" (unpublished paper, 1982) 17.

39. H. Peach and N. Carrington, eds., *The Face of the Land* (London: Design Industries Association, 1929), 8.

40. Ibid., 8.

41. T. Hennell, *Change on the Farm* (Cambridge: Cambridge University Press, 1936).

42. H. J. Massingham was widely known through his work and was extraordinarily prolific, producing more than twenty books on country life by 1944. In that year he added Thompson to the list of authors, which included Shelley and Gilbert White, whose work he had edited for a new public eager to consume a rural "tradition."

NARRATING POLLUTION

8

Public Perceptions of Smoke Pollution in Victorian Manchester

STEPHEN MOSLEY

Unlike many of today's environmental dilemmas, such as climate change and the thinning of the ozone layer, the smoke of Victorian Manchester did not elude the sensory perceptions of contemporaries. Coal smoke characterized the nineteenth-century urban atmosphere and affected the lives of all city dwellers, rich and poor alike. People lived and worked beneath lowering coal-black skies and imbibed the sulphurous, smoke-filled air with every breath they took. At the inaugural meeting of the Manchester Association for the Prevention of Smoke (MAPS) on 26 May 1842, the Reverend John Molesworth, the association's chairman, vehemently denounced a nuisance that "polluted our garments and persons" and that all the town's inhabitants "saw, tasted, and felt."[1] Despite the tangible nature of this form of air pollution, however, most contemporaries endured living in the midst of the city's "eternal smoke-cloud" without much outward sign of complaint. No popular mass movement against smoke developed in the city during the nineteenth century, even though its damaging effects *were* widely recognized.

How can we account for this seeming indifference toward smoke pollution? In this essay, I examine the dominant images and narratives that gave meaning to and created common understandings of a concrete en-

I am indebted to Greg Myers, Paolo Palladino, Thomas Rohkrämer, and John Walton for their helpful comments on a draft of this essay.

vironmental pollution issue in Victorian Manchester, the "smoke nuisance." Urban environmental degradation was rationalized and naturalized by the stories contemporaries told about air pollution, and, as William Cronon has recently argued, "to recover the narratives people tell themselves about the meanings of their lives is to learn a great deal about their past actions and about the way they *understand* those actions. Stripped of the story, we lose track of understanding itself."[2]

Environmental discourse about smoke in Manchester, England, was a bewildering stream of contested and contradictory claims and concerns. By analyzing how a variety of actors framed the phenomenon and investigating the context in which stories about the city's smoke unfolded, we can enrich our insights into how people defined, thought, and made choices about the local environmental conditions in which they lived. Thus far Victorian urban dwellers have been portrayed mainly as being uninterested in environmental issues. However, as I shall show, the citizens of nineteenth-century Manchester were much more than apathetic spectators where smoke pollution was concerned. To bring the main story lines about smoke pollution into sharper relief, I shall draw on a diverse range of texts, from newspaper stories, novels, and working-class autobiographies to postcards, poems, and popular songs. After briefly sketching the background to the problem, I focus on a "wealth" story line, assembling the components of a narrative that consistently emphasized the "inevitable" correlation between smoke, well-being, and economic prosperity. I then knit together the threads of a narrative that accentuated "waste," constantly stressing the unnecessary peril to the health of the urban workforce, the damage to the natural and built environment, and the uneconomic and willful misuse of Britain's finite natural resources. Finally, I suggest reasons why the concept of smoke control did not readily capture the public's imagination.

In the nineteenth century, Manchester was one of the world's most important cities as a result of the success of its cotton trade and associated industries. The first real industrial city, Manchester attracted visitors from all parts of the globe to wonder at the new "cityscape" of massive textile factories and warehouses and its forest of smoking chimneys. In the early 1780s the predominantly verdant and countrified town of Manchester had boasted just one solitary tall industrial chimney; by the early 1840s, the "shock city" of the Industrial Revolution had sprouted some five hundred factory chimneys, growing to around twelve hundred chimneys by 1898.[3] Coal consumption in the city had also increased substantially, from around 737,000 tons per year in 1834 to more than 3 million tons a year by 1876.[4] With people attracted by its mushrooming industries, the population of Manchester grew apace, increasing from around 40,000 to more than 76,000 inhabitants between 1780 and the first census of 1801. Sustained, dynamic urban growth saw Manchester's population more than treble to reach 242,000 in 1841, and only thirty years later the city had a population of some 351,000.[5] The sight of black sulphurous smoke billowing from the explosive growth of new factories and domestic hearths prompted Leon Faucher to compare Manchester to an active volcano in the 1840s, and Major General Sir Charles Napier, appointed commander of the troops of the Northern District in 1839, called Manchester "the chimney of the world . . . the entrance to hell realised."[6] From the turn of the nineteenth century the city's ever-deepening smoke cloud was a constant element of the urban environment. By the 1880s, after a century of rapid urban and industrial growth, Manchester, once "the symbol of a new age," had come to epitomize the smoke-begrimed, polluted industrial city. However, at the same time a positive, utilitarian image of Manchester's blackened physical environment had evolved, drawing on cultural values and beliefs that reflected its citizens' definition of themselves as an urban industrial workforce. For the steam-powered mills had brought material wealth for many of the city's inhabitants as well as environmental problems.

The first of the story lines that dominated the public's understanding of the production of smoke in Victorian Manchester contended that a factory chimney and, for that matter, a domestic chimney belching out black smoke symbolized the creation of wealth and personal well-being. Most of Manchester's manufacturers, its magistrates and councilors, members of its trade associations and chamber of commerce (with two or more of these positions of authority often held by one and the same person), and its substantial workforce seemed to subscribe wholeheartedly to this narrative. The smoke was represented as a necessary by-product of industry: "the inevitable and innocuous accompaniment of the meritorious act of manufacturing."[7] The production of smoke warranted no apologies from most industrialists, who pointed to their smoking chimneys as a barometer of economic success and social progress.

The city's booming industries, especially cotton, provided numerous job opportunities and produced rising living standards for an ever-increasing number of Manchester's working class—particularly after 1850. Nevertheless there were periodic slumps in the cotton trade, most notably the trade depression of 1837–1843, the Cotton Famine of the early 1860s, and the cyclical pattern of slumps known as the "Great Depression" of the 1870s to 1890s, as well as the challenge of German and American competition looming on the horizon. Traveling around the cotton towns of Lancashire during the very lean year of 1842 and finding the factories of Bolton, near Manchester, hard at work, William Cooke Taylor of Trinity College, Dublin, exclaimed, "Thank God, smoke is rising from the lofty chimneys of most of them! for I have not travelled thus far without learning, by many a painful illustration, that the absence of smoke from the factory-chimney indicates the quenching of the fire on many a domestic hearth, want of employment to many a willing labourer, and a want of bread to many an honest family."[8] The image of thousands of smokeless chimneys, as envisioned by the smoke reformers of the period, was almost certain to cause alarm and anxiety among Manchester's working classes. Concerns about the absence of smoke in the industrial city found expression in popular culture's representations of the dilemma. During the Cotton Famine, a cyclical slump exacerbated by the American Civil War, a poem entitled *The Smokeless Chimney* sold well, chiefly at Britain's railway stations, in aid of the Relief Fund for Lancashire's unemployed textile workers. Written in 1862 by Mrs. E. J. Bellasis, under the pseudonym of "A Lancashire Lady," it mirrors Taylor's earlier personal narration of the meaning of smoke:

Traveller on the Northern Railway!
Look and learn, as on you speed;
See the hundred smokeless chimneys,
Learn their tale of cheerless need.

"How much prettier is this country!"
Says the careless passer-by.
"Clouds of smoke we see no longer.
What's the reason?—Tell me why.

"Better far it were, most surely,
Never more such clouds to see,

Bringing taint o'er nature's beauty,
With their foul obscurity."

Thoughtless fair one! from yon chimney
Floats the golden breath of life.
Stop that current at your pleasure!
Stop! and starve the child,—the wife!

Ah! to them each smokeless chimney
Is a signal of despair.
They see hunger, sickness, ruin,
Written in that pure, bright air.

"Mother! mother! see! 'twas truly
Said last week the mill would stop!
Mark yon chimney,—nought is going,—
There's no smoke from 'out o'th top!'

Weeks roll on, and still yon chimney
Gives of better times no sign;
Men by thousands cry for labour,—
Daily cry, and daily pine.

Let no more the smokeless chimneys
Draw from you one word of praise.
Think, oh, think! upon the thousands
Who are moaning out their days.

Rather pray that, Peace soon bringing
Work and plenty in her train,
We may see these smokeless chimneys
Blackening all the land again.[9]

Bellasis's paean to air pollution (here much abridged) contains many of the intrinsic cultural messages that were essential for the propagation of the myth that smoke was inextricably linked to health, happiness, and prosperity. Workers wait despondently for smoke to issue from the lifeless factory chimneys; and it is smoke, the "golden breath of life," and not clean air, that would bring the urban masses employment, comfort, and plenty.

The importance of coal and the cotton textile industry for growth and prosperity in Manchester was also widely recognized in the lyrics of the popular songs of the period. In the 1840s and 1850s, for example, the comic song *Manchester's Improving Daily,* composed by Richard Baines, became a great favorite with the city's working inhabitants. The first verse went as follows:

> In Manchester, this famous town,
> What great improvements have been made, sirs;
> In fifty years 'tis mighty grown,
> All owing to success in trade, sirs;
> For we see what mighty buildings rising,
> To all beholders how surprising;
> The plough and harrow are now forgot, sirs,
> 'Tis coals and cotton boil the pot, sirs.
> Sing Ned, sing Joe, and Frank so gaily,
> Manchester's improving daily.[10]

While coals and cotton provided the workers' daily bread, this act was not usually accomplished without the "inevitable" production of large volumes of black smoke from the city's industrial chimneys. This view of coal and smoke was widespread and is reproduced in the following verse about Glasgow:

> There's coal underground,
> There's coal in the air,
> There's coal in folk's faces,
> There's coal—everywhere;
> But—there's money in Glasgow![11]

In an "age of smoke," popular poems and songs generated associations that helped naturalize and rationalize the relationship between wealth and air pollution in the industrial towns and cities of Britain. By the end of the nineteenth century, the image of a smoking factory chimney had become indivisible from employment and a full stomach in the minds of the urban masses.

On a more modest scale, a generous amount of smoke seen freely issuing from any one of the city's many thousands of domestic chimneys signified a working family's continued good fortune. Mrs. A. Romley Wright, who taught domestic economy classes in Manchester, illustrates the symbolic power of the smoke emitted from "the popular British institution" of the open coal fire: "The kitchen fireplace is filled with coal—large pieces, of course, for roasting. A volume of smoke rushes up the chimney, and the admiring neighbours may ejaculate 'Oh, *what* a dinner Mr. so and so must be having.'"[12] An extravagantly smoking chimney pot visibly demonstrated to onlookers that a family was doing well economically, and might even have enhanced their social status in the community. The smokeless fuel coke, although relatively inexpensive, was unpopular among the city's inhabitants. Coke did not make a good blaze in the hearth and was widely perceived as "a fuel of poverty."[13]

Domestic life revolved around the fireplace, especially during the cold, damp, and dreary winter months. A blazing fire imparted much more than an agreeable degree of heat. The domestic hearth was closely associated with the notion of human warmth, signifying love, friendliness, and a sympathetic, comfortable environment. There are innumerable popular images, both visual and literary, extolling the pleasures of hearth and home that date from the Victorian period. The popular culture of the day often depicts a family and friends seated around a roaring fire, swapping stories, enjoying eating and drinking together, singing songs, reading aloud, or simply watching the shapes made by the flames. A verse from the Lancashire dialect writer Edwin Waugh's short poem *Toddlin' Whoam* encapsulates the powerful attractions of the domestic hearth:

> Toddlin' whoam, for th' fireside bliss,
> Toddlin' whoam, for th' childer's kiss;
> God bless yon bit o' curlin' smooke;
> God bless yon cosy chimbley nook!
> I'm fain to be toddlin' whoam.[14]

Although such representations of hearth and home were often overly romanticized, it would be extremely pessimistic to suppose that such pleasant activities were not experienced by most working-class families, at least from time to time. Smoke spouting from chimney pots and cheerful open coal fires were, then, symbols that were commonly employed by contemporaries to indicate that times were relatively good. The other side of the coin—the cold, fireless grate—was an image that was used by novelists of the period, from national figures such as Charlotte Brontë, Mrs. Gaskell, and Charles Dickens to local working-class dialect authors such as Waugh and Ben Brierley, to denote want and poverty. And when, for example, Waugh wrote of "fireless hearths, an' cupboards bare" in the song *Hard Weather* (penned during the acute recession of 1878–1879), he would without question have sent a pang of anxious recognition through many of those who heard or sang it.[15]

The existence of bad trade conditions was not a prerequisite for the success of the story line that smoke was inevitable and denoted economic prosperity. At midcentury, for example, a time of neither boom nor bust in the city, the journalist Angus Reach wrote in the *Morning Chronicle,* "Purify the air of Manchester by quenching its furnaces, and you simply stop the dinners of the inhabitants. The grim machine must either go on, or hundreds of thousands must starve."[16] This message was repeated unremittingly by Manchester's employers, who were increasingly worried about the squeeze on profit margins and market share in the face of foreign competition. Their views, closely associating smoke with continued economic growth, were regularly reported at length in the *Manchester Guardian* and other local newspapers. Reginald Le Neve Foster, an influential director of Manchester's Chamber of Commerce, countered one of the City Council's many attempts to enforce the law against smoke pollution by declaring that if it succeeded, "they would drive away all their industries, . . . and Manchester would soon become one of the 'dead cities' of the world."[17] This was a narrative that was to a large extent shamelessly predicated on negative images of smokeless chimneys. It played constantly and effectively on immediately intelligible fears about what life in the industrial city would be like *without* its familiar and reassuring smoke cloud. Just as today, the issue of pollution control was often viewed in simplistic terms, with the manufacturers presenting a stark choice between smoky prosperity or economic stagnation if environmental safeguards were proposed.

The Victorian's well-documented abhorrence of dirt and filth did not always extend to coal smoke, which was often portrayed as good, honest dirt and not as "matter out of place."[18] Indeed, that Manchester's smoking chimneys came to be widely interpreted as benign signs of progress and prosperity is also indicated by a northern expression that has survived to this day: "Where there's muck, there's brass." As a result, the unsightly black face that the hard-working city of Manchester presented to the world could be viewed uncritically by some contemporaries. In 1887, for example, a contributor to the *Manchester Guardian* wrote,

Figure 8.1 Beautiful Manchester, early twentieth century. Postcard in the collection of Stephen Mosley.

"Physically, we must admit that Manchester does not make a good show, except of dirt; but it is only work-day dirt after all—the grime of a collier who has to deal with coal, the dust of a miller who has to apologise for his floury proportions."[19] Smoke was represented as beneficial and innocuous, a form of dirt that constituted no great threat to life and health. By the turn of the twentieth century, affectionate, tongue-in-cheek images of the city's smoking factory chimneys were appearing on postcards bearing the legend "Beautiful Manchester" (Figure 8.1). The image of a smoking industrial chimney had become as comforting psychologically as that of the domestic hearth. The production of smoke was commonly understood and hailed as an infallible sign that Manchester was a flourishing and enterprising city. Throughout the nineteenth century and beyond, the story line continued to resonate with meaning, with the local businessman Reuben Spencer writing of Manchester in 1897, "The factories are still there, the 'incense of industry' still floats in clouds above the tops of countless towering chimneys, the throngs of busy workers are more numerous than ever, and the whirr of gearing and the hum of machinery resounds in a hundred streets, whence emanate a thousand different wares for the use and benefit of the peoples of the earth."[20]

As late as 1913 *Black's Guide to Manchester* told of the city's "thick cloud of smoke that turns to invisible gold."[21] The narrative of wealth was imparted through a great variety of texts in Victorian Manchester, all of which bracketed smoking chimneys with healthy trade conditions, stable or rising living standards, and personal well-being. Charles Dickens sardonically captures the spirit of the age when he has Josiah Bounderby of Coketown say, "First of all, you see our smoke. That's meat and drink to us. It's the healthiest thing in the world in all respects, and particularly for the lungs."[22] But despite the willingness of the majority of Manchester's manufacturers, politicians, and workers to endure polluted air in the name of growth and prosperity, an active minority of influential reformers questioned the popular belief that smoke was synonymous with economic and social progress and countered with a compelling story line of their own.

The second of the story lines that conferred symbolic meaning to the production of smoke reflected the values and beliefs of a largely middle-class, educated, and professional elite, who, rather than viewing smoke as signifying prosperity and progress, saw the columns of sulphurous black smoke as "barbarous" signs of waste and inefficiency. Doctors, scientists, lawyers, clerics, architects, and others from the burgeoning professional ranks, along with several of Manchester's leading merchants and manufacturers, all promoted this skeptical alternative narrative. From the 1840s on, many reformers banded together to form antismoke societies in the city, among which were MAPS and the Manchester and Salford Noxious Vapours Abatement Association (NVAA), founded in 1876. The antipollution activists challenged entrenched cultural values and beliefs about Manchester's "productive" smoke by holding public meetings against air pollution; by regularly inviting leading "experts" to lecture on the subject; by testing and exhibiting smoke abatement technology; and by publishing articles and letters in newspapers, magazines, and journals.[23] Coal smoke, according to this story line, meant a failure to make profitable use of valuable and finite natural resources and a reckless waste of irrecoverable energy. Smoke meant the needless defacement and destruction of the city's buildings and green spaces; it caused a needless and preventable loss of life and health; and, finally, it represented a serious threat to Manchester, Britain, and Empire. The narrative that smoke was synonymous with waste was a denser, more complex response to the dilemma posed by smoke pollution and was conveyed in considerably fewer, and often less accessible, texts. However, although much of what follows was reconstructed from sources that did not enjoy an extensive popular readership, these narratives *were* widely disseminated in both the local and national presses.

The towering factory chimneys of the new industrial towns conveyed coal smoke quickly and cheaply into the "vast atmospheric ocean" overhead, where it was thought the unburnt products of combustion would be harmlessly diluted and dispersed to a "safe distance" from urban areas.[24] However, as the number of smoky industrial towns mushroomed in southeast Lancashire and elsewhere, this system of pollution removal had the unwanted effect of displacing the problem from one municipality to the next. An appeal to economic rationality was the initial response to this galling situation, with smoke reformers portraying the destructive results of air pollution as nothing less than a burdensome local tax.[25] In 1842 the manufacturer Henry Houldsworth, a prominent member of MAPS, calculated the financial cost to Manchester's inhabitants in "washing, cleansing, and keeping clean persons, garments, furniture and houses" to be "not less than £100,000 per annum"—undoubtedly one of the first of the many attempts by Victorian environmental reformers that used cost-benefit analysis in trying to persuade people to reduce air pollution in urban areas.[26] The meteorologist Rollo Russell listed some twenty-four different forms of loss or damage caused by smoke in London in his cost-benefit exercise of 1889, including the extra gas used for lighting all year round because of the loss of sunlight; reduced capacity for work due to ill health; and the destruction of trees, plants, shrubs, flowers, vegetables, and fruits. Although he did not include "uncertain items," such as the effects of the residence outside the metropolis by all those wealthy enough to avoid the smoke, the total cost still reached £5,200,500.[27] By the early years of the twentieth century, Manchester's Air Pollution Advisory Board reported that the city's smoke was costing its householders no less than £242,705 per annum in extra washing alone. The report continued, "Not only does black smoke mean waste in itself, but it causes further waste. Everybody knows how much it disfigures, but it is not generally known how much it destroys. It levies what may be called the Black Smoke Tax, and everybody living in Manchester pays this tax . . . Black smoke means not only an aesthetic but also an economic loss."[28]

Coal smoke was also depicted as the visible failure of manufacturers to capitalize economically on the nation's coal stocks by burning their fuel efficiently. Smoke control was ceaselessly posited as a valuable business proposition for industrialists, with Manchester's antismoke campaigners claiming that improved, mechanized fuel technology or even "ordinary care" taken in the operation of hand-fired furnaces could mean substantial financial savings. Clouds of black smoke, according to the narrative of waste, denoted nothing less than pounds sterling hemorrhaging needlessly from Manchester's chimneys into the skies above. At midcentury the smoke-filled urban air was depicted as a vast, unused "aërial coalfield" by the *Times*.[29]

Economic concerns over wasted fuel were augmented after midcentury by severe criticisms of the irresponsible depletion of limited natural resources. The noted sanitary reformer Dr. Neil Arnott mournfully denounced the thriftless squandering of Britain's coal reserves: "Coal is a part of our national wealth, of which, whatever is once used can never, like corn or any produce of industry, be renewed or replaced . . . To consume coal wastefully or unnecessarily, then, is not merely improvidence, but is a serious crime committed against future generations."[30] In 1850, the German physicist Rudolph Clausius had formulated the second law of thermodynamics, developing the concept that all energy becomes disorganized and dissipates over time, eventually being lost forever.[31] As ideas concerning entropy filtered down to the public domain, industrialists, who treated the nation's coal reserves as if they were an *inexhaustible* asset, were constantly urged by reformers to conserve what were now recognized to be *finite* fuel reserves. Nor did the inefficiency of the smoky domestic hearth escape censure by contemporaries, with John Percy estimating in 1866 that "in common domestic fires . . . seven-eighths, and even more, of the heat capable of being evolved from the coal pass up the chimney unapplied."[32] Russell went so far as to advocate that local authorities should levy a punitive tax on the open kitchen range of the "wasteful householder."[33] By the early 1860s serious concerns were being voiced as to just how long coal supplies in Britain would last, with Sir William Armstrong, in his presidential address to the British Association, calculating that stocks would last only for another 212 years. In 1865, the economist W. Stanley Jevons heightened these concerns when he estimated that Britain's coal reserves would be exhausted in no more than a century.[34] The question of a threat to Britain's future power and prosperity was frequently highlighted, and featured prominently in Jevons's work. "When our main-spring is here run down, our fires burnt out," he wrote, "Britain may contract to her former littleness, and her people be again distinguished for homely and hardy virtues . . . rather than for brilliant accomplishments and indomitable power."[35] Without the conservation of coal—the source of *all* power—it was argued, Britain and Manchester had nothing to look forward to but a gradual decline into mediocrity.

Ideas that connected smoke with wasted resources were common in the popular press of Britain in the last quarter of the century. In 1889 a *Manchester Guardian* account of a crowded public meeting against smoke pollution held at Manchester Town Hall voiced the main themes of this strand of the story line. Lord Egerton of Tatton, who chaired the meeting, is reported as declaring that the "consumption of smoke" would result in "pecuniary gain" to the city's manufacturers, who needed to be "enlighten[ed] . . . as to their own interests, to show them that the emission of smoke into the air was a waste of valuable carbon which ought to be in the furnace." Sir W. H. Houldsworth, M. P. for Manchester N. W., proclaimed to the large and influential gathering that "he himself believed that the consumption and prevention of smoke was economical . . . because the black smoke which they wanted to stop was actual force and energy going into the air."[36] Air pollution was increasingly vilified as "barbarous and unscientific," and readers were regularly bombarded by representations of the smoking chimney as indicative of wasted money, energy, and natural resources.

The damage smoke caused to the health of Manchester's inhabitants was an important component of the narrative of waste and one that acquired an increasingly apocalyptic edge as the century progressed. The detrimental effects of local smoke on the human respiratory system had been pointed out as early as 1659 by John Evelyn in his pamphlet *A character of England,* where he observed that "pestilential smoak . . . fatally seiz[es] on the lungs of the inhabitants [of London], so that the cough and the consumption spares no man."[37] However, at Manchester concerns about the smoky atmospheric conditions causing ill health did not surface conspicuously until 1842, when MAPS was formed. The Reverend John Molesworth, for example, told the Select Committee on Smoke Prevention of 1843 that the smoke-drenched air of the city was "no doubt unhealthy" and "must tend to disease."[38] The smoke abatement movement emerged at a time when the "condition of England" question was attracting great attention, with public health reformers gathering damning statistical evidence regarding the dangers of unhealthy water supplies and inadequate drainage and sewer systems in Britain's towns and cities.[39] Numerical data were used to link air pollution to the increased incidence of chronic respiratory diseases, especially bronchitis, in poor urban areas, with one report showing that the average death rate from respiratory diseases between the years 1868–1873 was just 2.27 per thousand in Westmoreland, 3.54 per thousand for the whole of England and Wales, 5.12 per thousand for Salford, and 6.10 per thousand for Manchester. In 1874 the death rate from these causes had risen to 7.70 per thousand in Manchester, leading the report's author to conclude "that Manchester suffers more from diseases of respiratory organs than any town or city in England."[40] In 1882 Dr. Arthur Ransome, president of the Manchester and Salford Sanitary Association, calculated that in the preceding decade some 34,000 people had died from "diseases of the lungs" in Manchester and Salford. Ransome argued that smoke pollution was a significant—and preventable—cause of this mortality and that its abatement "would save many useful lives in the next generation."[41] Concerns about the health risks posed by smoke were often expressed in the same terms as those that communicated the negligent misuse of mineral resources.

Then, as now, reformers consistently used the burgeoning death rates from respiratory diseases to keep the issue of air pollution in the public eye. But while considerable progress was made in improving sanitation and lowering death rates from "dirt diseases" such as cholera and typhoid during the nineteenth century, mortality statistics could not conclusively prove the relationship between ill health and the burgeoning smoke cloud. Increasing death rates from respiratory diseases were also associated with other environmental sources of illness, such as damp, overcrowded housing. Frustration at the ineffectiveness of "reform by numbers" in the case of air pollution saw a growing number of impassioned appeals to the public's moral sensibilities, with, for example, the socialist Edward Carpenter writing emotively,

> any one who has witnessed . . . the smoke resting over such towns as Sheffield or Manchester on a calm fine day—the hideous black impenetrable cloud blotting out the sunlight, in which the very birds cease to sing,—will have wondered how it was possible for human beings to live under such conditions. It is probable, in fact, that they do not live . . . The workers, producers of the nation's riches, dying by thousands and thousands, choked in the reek of their own toil.[42]

By the last quarter of the century, the notion that smoke was contributing markedly to ill health and mortality in the city had reached beyond medical circles to acquire a wider resonance. The idea had become an important part of public discourse concerning urban air pollution, as this letter to the *Manchester Guardian* illustrates:

> The great city of Manchester stands in the unenviable position of being one of the unhealthiest cities in the kingdom, with an appalling death-rate behind it. The evils are not far to seek. They are, in my opinion, mainly due to . . . the smoke demon and noxious vapours. So much has been written and said on this nuisance that I can add nothing fresh, only to hope that they who are charged with guarding the public health will be alive to their duties and enforce their power to abate the evil, which is killing downright our boys and girls—the men and women of the future—and let them breathe pure air and not poison . . . it is high time that something practical were done to make it what it ought to be—the second city in the Empire, healthy and cheerful, instead of what it is, insanitary and gloomy.[43]

The smoke pollution that veiled the city, rather than being a cause for celebration, was represented to the public as nothing less than a funereal pall (Figure 8.2). By the latter decades of the century, however, concerns about the putative effects of smoke on health were not limited solely to an increase in respiratory diseases.

The smoke-laden atmosphere of the city obscured the sun, and this "destruction of daylight" led to fears that smoke pollution was contributing to the general physical and moral deterioration of the urban working populace—a much debated concern throughout the century. As early as 1876, Dr. Thomas Andrews, president of the prestigious British Association for the Advancement of Science, stated, "There can be no doubt that the prevalence of smoke in the atmosphere of our large towns tends to deteriorate the physical condition of our people." Arthur Ransome argued in 1882 that the absence of sunlight in urban areas was contributing to "the pallid and unhealthy and stunted appearance of our town populations." In an article entitled *"Smoke, and its Effects on Health,"* the anonymous *"Lucretia"* discussed yet another dimension to

FIGURE 8.2 The smoke demon, 1893. From Robert Barr, "The Doom of London," *The Idler* 2 (1893): 7.

the narrative: that the degeneration of the urban populace, caused directly by Manchester's smoking chimneys, threatened the very survival of the city itself:

> A stunted, scrofulous, and ricketty working population has been raised up in our midst . . . The present existing facts of physical and moral deterioration of the human type which apparently lives, eats, and has its being amongst the lower classes of this large city, certainly shakes to its foundation Darwin's theory of the "survival of the fittest" . . . the smoky atmosphere of Manchester is not *all* that can be desired . . . The clearing of the atmosphere is one of the greatest necessities of the age; and we should consider it so, . . . the proof of the great dangers existing to the prosperity of Manchester, socially, physically, and commercially, is the apathy of the general mass of its inhabitants with regard to the "smoke nuisance."[44]

The narrative also strongly emphasized the perceived moral degeneration of the urban masses. It was claimed that "hundreds of thousands of English children are now growing up into men and women . . . with no sign of a green field, with no knowledge of flowers or forest, the blue heavens themselves dirtied by soot—amid objects all mean and hideous, with no entertainment but the music hall, no pleasure but in the drink shop."[45] It was argued that to compensate for the grim local environmental conditions—the lack of daylight, natural color, and vegetation in the drab industrial city—many of the working classes simply wasted their scant financial resources on "vulgar" entertainments, gambling, and drinking excessively, which was, in the memorable phrase, "the shortest way out of Manchester."

This story line gained currency in 1899, when of eleven thousand men in Manchester who volunteered for military service in the Boer War, eight thousand "were found to be physically unfit to carry a rifle and stand the fatigues of discipline." Concern did not end there, however, as of the three thousand who were accepted to serve in the army, only twelve hundred were found to be even "moderately" fit.[46] Following the Boer War, during which Britain's urban-raised soldiers were thought to have performed poorly in comparison with their country-bred opponents, Fred Scott, secretary of the Manchester and Salford branch of the Smoke Abatement League, warned that the "smoke nuisance" was seriously undermining the health of the nation, "sapping the virility of our people—that grand heritage which has made the British the greatest colonising and conquering race the world has known."[47] The ethos of imperialism was based on virility, and smoke pollution was thought to seriously impair the development of "a manly, vigorous, enterprising, healthy race which will hold its own against all foreign competition."[48] The colonies were depicted by the reformers as healthy limbs on a decaying body, and the wisdom of the state in continuing to allow the air to be polluted by coal smoke was seriously questioned. From the 1880s on, the smoke reformers' narrative repeatedly emphasized the degeneration of the lower classes of the city, aiming to shake the public's confidence in a viable future for Manchester, Britain, and Empire. This grim vision of the future for the smoky city was intended to countervail the doom-laden story line that had been constructed for a smoke-free Manchester.

The skeptical story line concerning smoke production also vigorously attacked the serious damage being caused to Manchester's recently erected "architectural beauties" and the wholesale destruction of vegetation in the city. The 1840s and 1850s had seen the flowering of a competitive form of civic pride in Britain that produced in the second half of the century not only an abundance of monumental town halls, public libraries, and municipal art galleries, but also a large crop of new public parks where plants from every corner of a growing Empire could be displayed. Whether in art, architecture, parks, or ornamental gardens, the emphasis was on spectacle, and the lofty aspirations of Manchester's manufacturing and commercial middle class found their expression in public buildings such as Alfred Waterhouse's neo-Gothic town hall, which took nine years to construct at a cost of £1 million. However, the architecture of the city dubbed "the Florence of the North" by the architect Thomas Worthington, which was designed to reflect the sophistication, power, and status of its urban "aristocracy" ad infinitum, rapidly began to blacken, and its elaborately carved stonework to crumble, as a result of the corrosive action of the viscid, sulphurous black smoke. Furthermore, trees, shrubs, flowers, and grass all struggled for survival in Manchester's parks and gardens due to the problem of acid rain, and paintings were placed behind glass to protect them from the smoke-laden air. Discontent regarding the grimy condition of the city, and the harmful effects of smoke on the natural environment, was already evident in the early 1840s, when an editorial in the *Manchester Guardian* complained about the city's dying trees and shrubs and stated that smoke pollution "has for years been the standing reproach of the town."[49] The "unloveliness" of Victorian Manchester was a sensitive issue for many contemporaries, since from an aesthetic perspective it was thought to compare poorly to other great cities of the world, such as London, Paris, Rome, or indeed, Florence.

Discolored, decaying buildings and stunted vegetation were a recurring motif of the skeptical narrative regarding smoke, as the murky urban environmental conditions were thought to effectively advertise "the waste which goes along with these aesthetic and hygienic backslidings." If medical statistics were unable to prove the link between smoke and ill health, this component of the story line frequently highlighted the visibly damaging effects of smoke on the city's masonry and vegetation as important indicators of what air pollution might be doing to human health. The high acidity of the city's rain, a direct consequence of the ever-expanding smoke cloud, was eroding the headstones in Manchester's cemeteries. This led the Town Clerk of Manchester, Sir Joseph Heron, to ask, "If the gravestones were suffering, what would be the effects on those above?"[50] It was but a small step for some contemporaries to use imagery of deteriorating, etiolated vegetation as a commonsense analogy to illustrate what they thought was happening to the urban populace. Ernest Hart, chairman of the National Smoke Abatement Institution and editor of the *British Medical Journal,* took "very carefully" the counsel of the leading medical authorities of the day, including Sir Andrew Clark, Sir William Gull, Sir James Paget, Dr. Alfred Carpenter, and Manchester's Ransome, concerning the effects of smoke on health before giving the following evidence to the Select Committee investigating Smoke Nuisance Abatement in 1887:

> The increase in the volume of smoke emitted in London, within the last 20 years, has produced perceptible differences in the health of the people, differences which you can measure pretty accurately by certain biological standards. For instance, Mr. Stansfield was telling me that not very many years ago roses could be grown successfully . . . at Prince's Gate, but that now it is impossible to grow roses there . . . At the present moment the very last conifer, I believe, is dying, or dead, in Kensington Gardens . . . all the vital processes of healthy life are allowed to droop for lack of the

> actinic rays of the sun: they are struck down by the increasing volume of smoke, which means a general deterioration of the health of the children, and grown-up people also.[51]

The reformers asserted that if smoke was abated and something of nature restored to the city, the results would include a happier, healthier, and more civilized urban workforce; an increase in artistic creativity and industrial output; higher standards of production; and, ultimately, the maintenance of Manchester's and Britain's standing in the world. They continually stressed that an investment in the regeneration of the non-human environment—in clean air, open spaces, parks, and healthy, flourishing vegetation—was also a way to increase or measure the public wealth of a city. However, despite widespread and favorable press coverage, the narrative of "waste" failed to overturn the dominant cultural myth that smoke equaled prosperity.

By the 1880s, a century's experience of living with smoking chimneys had given a cultural permanence to the notion that coal smoke denoted wealth. The correlation between smoke and prosperity had become so deeply embedded in the culture of northern industrial society that many city dwellers did not often think to complain about the murky atmospheric conditions. The journalist and socialist Allen Clarke wrote of his working-class upbringing in Bolton: "Living there, I had grown familiar with its ugliness, and familiarity oftener breeds toleration than contempt; I had accepted the drab streets, the smoky skies, the foul river, the mass of mills, the sickly workers, as inevitable and usual—nay, natural, and did not notice them in any probing, critical way."[52] Indeed, the gloomy city conditions helped to form, and strengthen, the British urban dweller's legendary attachment to the bright and cheerful open coal fire. Professor William Bone wrote of the national preference for coal fires: "An Englishman, oppressed as he is [by] . . . dreary sunless skies . . . seeks relief in his home at nights by his radiant fireside, and disregards with characteristic disrespect the vapourings of scientific cranks who condemn it as wasteful."[53]

Coal was one of life's necessities as far as the working classes were concerned, and finding a lump of coal in the street was looked upon as a sure sign of good luck. Robert Roberts considered that obtaining food and warmth were the greatest worries of the slum dwellers of Salford and set down the views of a regular customer to his father's shop: "'A full belly and a warm backside', Mrs. Carey would announce, 'that's all our lot want! I got a sheep's head boiling on the hob and a hundredweight o' nuts [coal] in the backyard. What more could folks wish for in winter?"[54] The answer to Mrs. Carey's question is regular employment, and this is where the potency of the "wealth" story line really comes into its own. We must not forget that the only occasions on which urban dwellers had experienced clean city air were in times of hardship, such as a trade depression or a strike. Their encounters with a smoke-free urban environment had been uniformly wretched, and under these circumstances it is not difficult to understand why most of Manchester's citizens accepted the polluters' customary story line; they preferred to cling to what they knew to be true. When the city's workers saw a smoking chimney, they did see wealth being created, and smoke issuing freely from their domestic chimney did signify their prosperity. Although both main story lines were at times grossly exaggerated, they did carry many different kinds of truth concerning smoke pollution that the city's inhabitants could readily identify with. However, the narrative of "wealth" was by far the more credible in the eyes of the urban workforce, who knew, from bitter, lived experience, that a smokeless chimney signified enforced idleness, hunger, and poverty.

The views of the mass of workers have been largely absent from discussions about urban environmental conditions. Working people, however, did hold strong opinions about air pollution, and glimpses of their perceptions of the "smoke nuisance" in the industrial towns of northern Britain do occasionally come to light—and not only in popular poems and songs. In conclusion, I suggest that the workers of Manchester employed an evaluative hierarchy where these story lines were concerned. There is ample evidence to suggest that substantial numbers of working-class people did complain about the harmful effects of smoke pollution, with Captain A. W. Sleigh, former assistant commissioner of police at Manchester, telling the Select Committee on Smoke Prevention of 1843 that "the poor people themselves consider it a very great nuisance . . . The whole population do. Lord Ashley was at Manchester some time before I left there, and he did me the honour of asking me to go round with him, to look at the condition of the poor people, together with other matters. We visited all the localities minutely, and they all complained of the smoke."[55]

Ill health; the Sisyphean task of attempting to keep homes, furnishings, and clothes free from soot; and the destruction of vegetation in the barren, smoky city were all issues that attracted their criticism. But the working classes unquestionably afforded a higher priority to the manufacturers' claims that they would be worse off *without* the industrial smoke with which they had come to associate jobs and economic well-being. That there was little active support for the smoke abatement movement among the working classes can be exemplified by an account in the *Manchester Guardian* of a public meeting against smoke, convened by the NVAA, at Broughton Town Hall on 11 December 1882. The meeting was planned with a view to putting pressure on Salford Corporation to prevent "the needless emission of black smoke . . . in the neighbourhood." The *Manchester Guardian* reported that there "was a large attendance of working men" at the meeting, and their hostility to the aims of the reformers soon became apparent. Despite an earnest appeal as to the "vital importance to themselves and their families that their health be sustained in order that they might continue to earn that income upon which both they and their family depended," the NVAA's various arguments concerning the deleterious effects of smoke made little impression on the assembled throng. A resolution berating the Salford Corporation, on being put to the meeting, "was lost by a very large majority, only four people in the body of the room voting for it." Directly after the vote had been taken, Thomas Horsfall, on behalf of the NVAA, asked, "Whose voice was it that ordered that no hand be held up in favour of the resolution. Was it that of a manager or overlooker?" Horsfall was immediately shouted down by the angry crowd. An opposing speaker, George Jones, a member of the Health Committee of the Salford Corporation, found great favor with the audience when he protested that he "did not think there was much to complain about in Broughton, and he would be sorry to see a persecution commenced against the manufacturers, for if they were driven from the borough, where would the bread of the working man come from?"[56]

The reformers claimed later that "it was evident from the opening of the meeting that a large number of workmen who entered the hall . . . had been sent to defeat the objects of the Association."[57] It is impossible to know whether or not these workers had been coerced by their employers into attending the meeting at Broughton, but it is likely that any use of coercion was minimal, as many working men undoubtedly saw their own interests as being intimately linked to those of their employers in this respect.

No single view of air pollution dominated to the exclusion of all others. The story lines of "wealth" and "waste" are two sides of the same coin, with smoke simultaneously existing as both good *and* evil in the minds of contemporaries in urban industrial areas. But the strong foothold that the straightforward, cohesive narrative of "wealth" had obtained in popular culture gave it great influence and staying power. The more fragmentary and scientific story line of "waste" was not as readily intelligible to the working classes and failed to duplicate or seriously undermine the former's authority. Trust in the purveyors of these differing knowledge claims also influenced the ways people made decisions about smoke in their uncertain day-to-day lives. Against the backdrop of cyclical economic depressions, especially during the Great Depression years, it is likely that both employers and employees came to share a sense of increased vulnerability and feared change and the unknown. The growing foreign challenge to Britain's commercial supremacy also eroded confidence in Manchester's future as cotton prices fell and production expanded less rapidly.[58] "Sun doesn't pay hereabout," one worker stated flatly to a reformer in 1890. "More smoke more work hereabout, at least, that's wot my master says."[59] The working classes were suspicious of and did not support the reformers' questionable initiatives that might limit industrial growth and endanger their often precarious livelihoods.

Most of Manchester's citizens were not indifferent to smoke, which they had come to view primarily as the "incense of industry" and only to a subordinate degree as a symbol of waste. Despite the palpability of the coal smoke, this nineteenth-century environmental issue was no less socially constructed or complex than today's intangible air pollutants. The "waste" narrative attracted widespread public attention but failed to break the hold of the enduring cultural myth that smoke equaled prosperity. Had they been asked about their goals in life, most working people in Victorian Manchester would have had clear and definite answers: a blazing coal fire in the hearth, a good meal, and the aim of raising—or at least not worsening—their material standard of living. At the end of the nineteenth century, as at the beginning, people still looked to the city's thousands of chimneys to gauge the condition of their world.

NOTES

1. "Prevention or Abatement of Smoke? Public Meeting at the Victoria Gallery," *Manchester Guardian,* 28 May 1842.

2. William Cronon, "A Place for Stories: Nature, History, and Narrative," *Journal of American History* 78 (1992): 1369. Maarten Hajer's recent study of environmental discourse in Great Britain and The Netherlands (*The Politics of Environmental Discourse: Ecological Modernisation and the Policy Process* [Oxford: Clarendon Press, 1995]) was also an influence on this essay.

3. See Stephen Mosley, "The 'Smoke Nuisance' and Environmental Reformers in Late Victorian Manchester," *Manchester Region History Review* 10 (1996): 43.

4. Love and Barton, *Manchester As It Is* (Manchester: Love and Barton, 1839), 26; Robert Angus Smith, "What Amendments are required in the Legislation necessary to prevent the Evils arising from Noxious Vapours and Smoke?" *Transactions of the National Association for the Promotion of Social Science* (1876): 518.

5. See Alan Kidd, *Manchester* (Keele, England: Keele University Press, 1993).

6. Leon Faucher, *Manchester in 1844: Its present condition and future prospects* (Manchester, England: Abel Heywood, 1844), 16; Napier quoted in Steven Marcus, *Engels, Manchester, and the Working Class* (London: Weidenfield & Nicolson, 1974), 46.

7. Thomas C. Horsfall, "The Government of Manchester," *Transactions of the Manchester Statistical Society* (1895–96): 19.

8. William Cooke Taylor, *Notes of a Tour in the Manufacturing Districts of Lancashire* (London: Duncan & Malcolm, 1842), 22.

9. John Harland, ed., *Lancashire Lyrics: Modern Songs and Ballads of the County Palatine* (London: Whittaker, 1866), 289–292.

10. Richard Wright Procter, *Memorials of Manchester Streets* (Manchester, England: Thos. Sutcliffe, 1874), 40–42.

11. Peter Fyfe, "The Pollution of the Air: Its Causes, Effects, and Cure," in Smoke Abatement League of Great Britain, *Lectures Delivered in the Technical College, Glasgow 1910–1911* (Glasgow: Corporation of Glasgow, 1912), 77.

12. A. Romley Wright, "Cooking by Gas," *Exhibition Review,* no. 5 (April 1882): 3.

13. Robert Roberts, *A Ragged Schooling: Growing Up in the Classic Slum* (Manchester, England: Manchester University Press, 1976), 73.

14. George Milner, ed., *Poems and Songs by Edwin Waugh* (Manchester, England: John Heywood, n.d.), 34–35.

15. Ibid., 107–109.

16. J. Ginswick, ed., *Labour and the Poor in England and Wales, 1849–1851: The Letters to the Morning Chronicle from the Correspondents in the Manufacturing and Mining Districts, the Towns of Liverpool and Birmingham, and the*

Rural Districts, vol. 1, *Lancashire, Cheshire, Yorkshire* (London: Frank Cass, 1983), 5.

17. *Manchester Guardian,* 20 June 1891.

18. The work of Mary Douglas provides a good starting point concerning human understandings of dirt and disorder. See especially Mary Douglas, *Purity and Danger: An Analysis of Concepts of Pollution and Taboo* (London: Routledge & Kegan Paul, 1966).

19. "The Ugliness of Manchester," *Manchester Guardian,* 17 August 1887.

20. Reuben Spencer, *A Survey of the History, Commerce and Manufactures of Lancashire* (London: Biographical Publishing, 1897), 48.

21. *Black's Guide to Manchester* (London: A. & C. Black, 1913), 1.

22. Charles Dickens, *Hard Times* (Oxford: Oxford University Press, 1989), 166.

23. For a summary, see Mosley, "The 'Smoke Nuisance,' " 44–45.

24. See, for example, Peter Spence, *Coal, Smoke, and Sewage, Scientifically and Practically Considered* (Manchester, England: Cave & Sever, 1857), 21–22.

25. "The Smoke Nuisance," *Manchester Guardian,* 7 June 1843.

26. "Prevention or Abatement of Smoke? Public Meeting at the Victoria Gallery," *Manchester Guardian,* 28 May 1842.

27. Rollo Russell, *Smoke in Relation to Fogs in London* (London: National Smoke Abatement Institution, 1889), 22–26.

28. Manchester Air Pollution Advisory Board, *The Black Smoke Tax* (Manchester, England: Henry Blacklock, 1920), 1–2.

29. "London Smoke," *Times,* 2 January 1855.

30. Neil Arnott, "On a New Smoke-Consuming and Fuel-Saving Fire-Place," *Journal of the Society of Arts* 2 (1854): 428.

31. David Pepper, *Modern Environmentalism: An Introduction* (London: Routledge, 1996), 230–233.

32. John Percy, "Coal and Smoke," *Quarterly Review* 119 (1866): 451.

33. Russell, *Smoke,* 27.

34. Brian W. Clapp, *An Environmental History of Britain Since the Industrial Revolution* (London: Longman, 1994), 152–156.

35. W. Stanley Jevons, *The Coal Question,* 3d ed. (1906; reprint, New York: Augustus M. Kelley, 1965), 459.

36. "The Prevention of Smoke in Towns: Meeting in Manchester," *Manchester Guardian,* 9 November 1889.

37. John Evelyn, *Fumifugium* (1661; reprint, Exeter: Rota, 1976), prefatory notes.

38. *Parliamentary Papers* (House of Commons), 1843 (583) VII, qs. 680, 686.

39. See Anthony S. Wohl, *Endangered Lives: Public Health in Victorian Britain* (London: Methuen, 1984).

40. *Manchester and Salford Sanitary Association Annual Report 1876* (Manchester, England: Powlson & Sons, 1877), 9.

41. Arthur Ransome, "The Smoke Nuisance. (A Sanitarian's View.)," *Exhibition Review,* no. 1 (April 1882): 3.

42. Edward Carpenter, "The Smoke-Plague and its Remedy," *Macmillan's Magazine* 62 (1890): 204–206.

43. "The Manchester Death-Rate," *Manchester Guardian,* 25 June 1888.

44. Andrews quoted in Smith, "What Amendments," 537; Ransome, "The Smoke Nuisance," 3; Lucretia, "Smoke, and Its Effects on Health," *Exhibition Review,* no. 5 (April 1882): 2–3.

45. Fred Scott, "The Case for a Ministry of Health," *Transactions of the Manchester Statistical Society* (1902–3): 100.

46. Carl Chinn, *Poverty amidst Prosperity: The Urban Poor in England, 1834–1914* (Manchester, England: Manchester University Press, 1995), 114.

47. Scott, "The Case," 99.

48. Sir James Barr, "The Advantages, from a National Standpoint, of Compulsory Physical Training of the Youth of This Country," in *Manchester and Salford Sanitary Association Annual Report 1914* (Manchester, England: Sherratt & Hughes, 1915), 22.

49. "The Smoke Nuisance," *Manchester Guardian,* 28 May 1842.

50. "The Smoke Nuisance," *Manchester Guardian,* 19 September 1888, 3 November 1876.

51. *Parliamentary Papers* (House of Lords), 1887 (321) XII, q. 338.

52. Allen Clarke, *The Effects of the Factory System* (1899; reprint, Littleborough: George Kelsall, 1985), 38.

53. William A. Bone, *Coal and Health* (London: 1919), 15.

54. Roberts, *A Ragged Schooling,* 71.

55. *Parliamentary Papers* (House of Commons), 1843 (583) VII, qs. 1553, 1554.

56. "The Smoke Nuisance in Broughton: Lively Public Meeting," *Manchester Guardian,* 12 December 1882.

57. *Manchester and Salford Noxious Vapours Abatement Association Annual Report 1883,* in *Manchester and Salford Sanitary Association Annual Report 1883* (Manchester, England: John Heywood, 1884), 82.

58. John Walton, *Lancashire: A Social History, 1558–1939* (Manchester, England: Manchester University Press, 1987), chap. 10.

59. H. D. Rawnsley, "Sunlight or Smoke?" *Contemporary Review* 57 (1890): 523.

9

Narrating the Toxic Landscape in "Cancer Alley," Louisiana

BARBARA L. ALLEN

WE UNDERSTAND our lives through stories establishing us as part of some community, connected through memories with those things that touch on our deepest sense of who we are. These stories help us give shape to the world around us, give form to the landscapes we inhabit, and map our disparate past onto our current situations. Personal and collective landscape narratives can be an important part of a liberatory civic discourse facilitating social and environmental change, particularly in a region that is a patchwork of Superfund sites, hazardous waste processors, oil refineries, and petrochemical plants (Figure 9.1). This describes the landscape of my project, the parishes between New Orleans and Baton Rouge along the Mississippi River—the landscape many residents call "Cancer Alley."

The stories that people tell about places rely on their expression of their experience, which they construct within the normalizing discourses of their culture.[1] If we unveil experience and its context, it is less about origins and truth and more about *how* we tell our stories to explain those things that constitute evidence and knowledge in our world. In the case of the toxic landscape of Cancer Alley, the productive language-instrument is the residents' use of events to explain complex environmental phenomena toward the creation of a new body of evidence and knowledge about their homes. A focus on the resident's verbal testimony acknowledges that narrative strategies are a way to *know* about one's environment in a way that is different from corporatist or scientific descriptions. The key in community environmental transfor-

FIGURE 9.1 Local graveyard in "Cancer Alley," Louisiana, in the late 1990s. Photograph by Barbara Allen.

mation is to privilege equally all forms of knowing and understanding the landscape, so that a diversity of stakeholders may have a part in constructing its boundaries and uses through public mechanisms.

A new paradigm for understanding the dynamics of community environmentalism is emerging. Opposed to the modernist project of individual actors exercising their will and power on a self-defined problem are hybrid place-situated concepts and identities that act as catalysts initiating social and environmental change. As Donna Haraway argues, "The ecological object of knowledge can no longer rely on the silent structured division between nature and culture. The ecological object of knowledge also includes human history and actions."[2] Place-situated identities, however, are neither simple encyclopedic definitions of ethnicity nor individual actors/agents merely "speaking out" of their own volition. Instead, I am interested in analyzing collective narratives about the environment constructed within resident community groups. Stanley Fish's notion of interpretive communities, or groups of people who "share a way of organizing experience," including "categories of understanding and stipulations of relevance," depicts most clearly this type of language-centered identity. This approach to identity sees itself as an engine of change, as the individual and the group are in a constant flux of interpretation and reinterpretation. As Fish notes, "The community, in other words, is always engaged in doing work, the work of transforming the landscape into material for its own project."[3] This metaphorical statement has literal applications in the structure of public discourse about Cancer Alley, a landscape transformed through human discourse. Action is continually being redefined within the rhetoric and storytelling of competing and complementary groups. What is important is how environmental transformation is both activated and amplified within narrative communities.

Historically, the parishes along the Mississippi River between New Orleans and Baton Rouge can be called the Acadian Coast. Some of the first colonial occupants of the region were the French Acadians exiled from Nova Scotia in the mid-1700s. They were granted land that was divided in the French *arpent* system, allowing families to own narrow strips of property extending from the river across farmland to the inland bayous. To this day, the riverfront landscape remains predominantly divided into arpents.

After the Civil War, African Americans were granted land along the river. Documents suggest that the land was deeded to large extended family groups for the purpose of subsistence farming.[4] For example, one of the most coextensive of the grants was given in 1888 to newly freed slaves in tracts ranging from 40 to 160 acres in the area now known as Geismar in Ascension Parish.[5] Today, together with the neighboring towns of Carville and St. Gabriel, Geismar has become one of the most polluted places in the United States and one of the major focuses of the environmental justice movement.

Until the middle of the twentieth century, this land along the river was used primarily for farming. The former plantation properties were planted with sugar cane, and the smaller landowners had subsistence crops and sometimes truck farms. Raising sugar cane as a cash crop, once a labor-intensive venture, had become highly mechanized by this time, forcing laborers into the cities in order to survive. Many of the remaining families were landowners, small and large, black and white, who were either unable or unwilling to move from their homes to better fortune. This was the economic landscape that the petrochemical firms encountered when they began the large-scale development of this region in the mid-twentieth century.

There are now eighteen petrochemical plants discharging 196 million pounds of pollutants annually between the two towns of St. Gabriel and Geismar, which are less than ten miles apart.[6] Sugar cane cultivation still takes place on the marginal lands too wet or inaccessible for industry. Often the purpose of planting cane is to accrue tax write-offs for the industry-owned property not used by the chemical plants.[7] This brief historic narrative of the shift from cane to chemicals in the region is continually referenced by the residents as they talk about the issues they now confront in this heavily polluted place.

My analysis is based on more than thirty hours of public environmental hearing testimony by the residents of the Mississippi River industrial corridor and on personal interviews with more than forty residents and state officials. Some of the narratives are from three environmental justice hearings held by the Louisiana Department of Environmental Quality (DEQ) in 1994, and others are from an array of hearings on air, land, or water quality since 1991.[8] The narratives function as both story and rhetoric, identifying the various stakeholders as part of an interpretive community. The environmental hearings were mandated by the state and federal governments after the substantial popular press on the region in the years after the United Church of Christ's Commission for Racial Justice reported on toxics in minority communities.[9] The majority of people who testified told complex and disturbing stories about how they perceived and reacted to the effects of the petrochemical industry on their residential environments. The hearing format did have some substantial performative differences from other less formal communicative formats; however, their overall language and narrative strategies were quite similar.[10] Additional one-on-one interviews conducted in 1997 augmented narratives not represented in the various formal hearings.

Every landscape can be described on many levels. For example, it can be read like a text (semiotics), narrated like a story (experience), analyzed in a materialist framework, or evaluated from a biological standpoint.[11] The industrial landscape of south Louisiana is no exception. The towns, rural communities, and farms interspersed with approximately two-hundred chemical plants and petroleum refineries along the seventy-five-mile stretch of river between New Orleans and Baton Rouge can be described using any combination of the above methods.

During the past decade, the interpretations of the landscape by industry and residents have been highly contradictory. Why are their environmental stories so divergent? Who are the various claim makers, what are their claims, and how have they arisen? Through examination of the conflicts in the discursive systems and the multiple meanings in the concepts they deploy, an insight into the strategies of power in the chemical corridor can be accessed for community purposes.[12] In this essay, I concentrate primarily on the local residents, mentioning other stakeholders as needed for clarification.

The term "landscape" has many interpretations. Although it was first used in the English language in the late seventeenth century to refer to a type of scenic painting, it came to mean "an awareness of knowable space" in its first uses in colonial North America.[13] With the growth of industry in the United States, the term "landscape" began to include the social, cultural, and political aspects of a certain terrain. In vernacular terms, it can also mean a panoramic view of the outdoors, as impressions of landscapes owe much to their visual appearance and can be easily manipulated in this way.

The visual landscape of Cancer Alley can be understood in sociopolitical terms. For example, there are multiple ways in which the chemical corridor has been visibly represented and altered by the various stakeholders in the region. The sheer size of the chemical plants forms an imposing presence on the horizon, and other elements further their dominating profile. High transparent industrial fences with ominous warning signs typically divide a neighborhood from corporate territory. These visible markers, with their vivid inscriptions, act as constant reminders of the relative power wrought by the industries over the people in the area. In addition, marking the landscape in a maze of colored lines, thousands of pipelines along the river carry feedstock and waste products throughout the area. Many of these pipelines transport dangerous materials, often under pressure, and are sometimes clearly labeled as such as they travel through the air and on the ground within the domain of the townspeople. Said Margie Richard, a resident of Norco, "We read the pipelines in front of our street. Every one of them states, 'Highly dangerous gases. In case of emergency, call Houston.'" Houston is more than three hundred miles away. Industry adversely affects the day-to-day lives of its neighbors in other ways. "There used to be a time we hung our sheets and stuff outside," said resident Debra Ramirez. "They would have all kind of black specks on them. We'd wipe our windows, just all kind of black stuff come out, and if that stuff was on your windows and your clothes . . . it was in your lungs as well."[14]

Neighbors claim that the chemical plants take advantage of the cloak of darkness, allowing much greater quantities of waste to escape from their stacks at night. Amos Favorite, a longtime resident of Geismar and a veteran, put it this way: "You ought to see this place at night . . . When these companies burn off their waste the air lights up like a battlefield. I'm telling you it's scary. Nighttime around here is like an evil dream."[15] Leroy Alfred, an elected official in the new town of St. Gabriel, concurred. "Sometimes here at night you can't even go outside—it's just that bad with emissions."[16]

Many of the trees in the industrial corridor show visible signs of distress. The top leaves are missing, which some residents attribute to the high concentration of noxious fumes from the nearby stacks. Sometimes the trees are dead in an entire area. Favorite said he can't grow healthy crops anymore because of air and water emissions. Many residents notice spots and shriveling on parts of their plants; sometimes they don't grow at all. The residents often wonder whether it is safe to eat the fruits and vegetables they do grow. Della Sullivan finally gave up on her houses near a chemical plant: "The paint gets off the houses, your roofs are ruined, your shrubbery is ruined, your garden is ruined, you couldn't have anything here. You're scared to let the children play outside."[17]

The daytime atmosphere of the toxic landscape, according to the residents, is also poisonous. Sometimes it is visible in the form of a dark cloud or a thin, low-lying fog. However, what is visible to the residents of Cancer Alley is often invisible to the company and state government officials. According to resident Sandra Ryan, "I would look out on rainy days as I sat in my wheelchair . . . the pollution was almost as black as these walls . . . I would call my sister, and she would call the Department of Environmental Quality, and by the time these folks would come out, well, they had to come at their convenience and when they'd come out it was no longer evident." Ramirez's experience is similar: "The DEQ comes out, whatever you smell or whatever you see, they don't see for some reason, and they made me feel like I was blind."[18]

Speaking historically is one way that Favorite claimed authority for his dystopian vision of the industrial landscape: "Looking back on history—and I've been here 67 years—I've never seen Geismar like it is . . . that these new industries that were moving here was impacting us with something that was killing us. Our water is all messed up, our air is messed up, and all our land is messed up."[19] Many community residents adopt a historical narrative that grants them the authority to speak as regional experts. Florence Robinson, a resident of the region and biology professor at Southern University, an African-American institution, testified that "most of the people of color communities in Louisiana that have been impacted by the petro-chemical industry were very old communities that preceded the industries." Rolinda Dorris confirmed this, stating that she, too, lives in a small, rural African-American community. "It was founded by ex-slaves in the late 1800s, and its existence is in jeopardy right now because of the industry that lines the Mississippi River."[20]

Many residents fear the very place they call home. The anguished townspeople tell stories of illness and disaster. One resident claimed that no one over the age of fifty lived in his neighborhood. Another talked of the number of cases of sarcoidosis, a rare lung disease, in her community. Richard told of a woman burned to death in a vinyl chloride fire in her own front yard. She described her street as the dividing line between a chemical plant and a petroleum refinery. She is worried about the proximity of the vinyl chloride pipes to her house and about the general safety of the petrochemical plants in her neighborhood. She explained that her sister had just died of sarcoidosis and four other people in her town have the same disease. "Sarcoidosis is riding the town like a ghost," she said. Sarcoidosis is a common illness among residents of the chemical corridor, but the medical establishment seems to know little about it except that it is fatal. She suggested that "Cancer Alley" is not appropriately descriptive; she calls her community simply "death alley."[21]

Naming is an important activity for citizens struggling to reclaim a healthy environment for their communities. This is evident in the fact that the names for the region vary dramatically, depending on the specific community. Some older members of the white community call it the Great River Road, referring to the days of the sugar plantation culture or its current status as a historical landscape attracting many cultural tourists visiting the numerous antebellum homes and outbuildings still extant. Cajuns sometimes call it the Second Acadian Coast, referring to it as an early site of French-Canadian refugee settlement in the eighteenth century. The government proudly refers to the region as the Industrial Corridor, marking its economic success in the scheme of southern development planning. Industrialists and company people often call it the Chemical Corridor, laying claim as the region's most powerful political constituent. Many current residents, however, call it Cancer Alley, referring to the multitude of health problems they face on a daily basis. People speaking at the environmental justice hearings or being interviewed could be readily identified with their political group simply by the name they used for the landscape.

More recently, the issue of naming has been central to the disagreements surrounding the siting of the Japanese-owned Shintech facility, the first new petrochemical plant to be built from the ground up in the region in almost twenty years. This new plant, if built, would have employed about 140 people and cost $700 million dollars to build. It would have been among the largest vinyl-chloride-producing plants in the world. Underground it would have been connected to the vast network of feedstocks and secondary products that runs along the Mississippi River corridor. The plant was to be located a few miles south of Geismar near the rural town of Convent. The proposed chemical plant's immediate neighbor was a small black community founded by freed slaves following the Civil War; the current residents call their community Freetown. Freetown does not appear on any maps and was not mentioned in any of the permit applications and documents originally filed with the DEQ. Later in the public hearing process, the oversight was first denied by officials, who said that there is no such town because it's not on the map. With citizen input, the community of Freetown came to be recognized as a legitimate rural settlement. It led one black employee of the DEQ to comment that the absence of the town's name on the map is only further evidence of racial bias evidenced in the state's policy and planning mechanisms.[22] In 1998, in the midst of the ensuing controversy over racial politics, Shintech withdrew its permit application to locate in Convent.[23]

Along with the visible signs, the maps, and the names, the landscape of this corridor is marked by audible signs of powerlessness. At intervals loudspeakers dot the telephone poles and other tall structures near the petrochemical facilities. They are sound alarms that notify the residents of a chemical leak, an accidental gas release, or an imminent explosion at one of the plants. Alarms are most likely to sound at night, when there are fewer people monitoring the plants and when the industry is both flaring and releasing waste products at a greater rate, according to nearby community members. Many people testified that they slept on the edge of their bed, often dreaming of alarm sounds.

Residents report spending many nights running through the fields with their children away from the alarms. Sometimes it would be hard on a moment's notice to tell which plant was sounding the alarm in order to know which way to run in the dead of night. At daybreak they would backtrack through the fields, locating the shoes, stuffed animals and other items children might have dropped while fleeing. Many families do not own a car, but a car would not help, anyway, as oftentimes the official evacuation route would be blocked by a train. The residents know the night landscape intimately on foot.

At a hearing, one woman spoke to a group of people at the back of the room. She was noticeably upset as she related her family's story of running through the night to escape the deadly gasses. One night the alarms woke them, and her mother and father were too feeble to run. She had to leave her parents in the house as she fled with her children. Soon after this experience, she abandoned her home and moved to another town. She does not understand why the companies have bought out her white neighbors but have left her small black community in terror.

Della Sullivan has had similar experiences. "We have run and run and run, and I run till I got tired of running and I decided to move out of the area," claimed Sullivan, who went on to testify that she has lost both her husband and her daughter to toxic chemicals. "I have been sick," she said. "The doctors can't hardly find out what's wrong."[24] The DEQ representative asked her how many people are left in her community. She replied that she is unsure, since so many people have died and she is unable to keep count. She fears her own backyard and thinks the place is killing her.

The residents of the former black community of Morrisonville, about seventy-five families, were so close to the Dow Chemical facility that the company eventually bought them out, beginning in the early 1990s. For years before then, Dow had installed "daisy boxes," or one-way radios, in their homes so that they could hear what was going on in the plant twenty-four hours a day. Instead of instilling confidence, these radios were perceived as invasive, and residents insist that Dow installed them for liability purposes so that the company officials could say the people knew what was going on in the plant, when in reality the radios provided little interpretable environmental information.

The corporate buyout of Morrisonville was one of several such purchases of African American communities during this period. The nearby communities of Reveilletown and Sunrise, both more than 100 years old, were razed by chemical companies concerned about contamination.[25] The above-ground inhabited landscape was not the citizens' only concern, however. In June 1989, the people of Morrisonville had been told that there was groundwater contamination under one of the vinyl chloride units about 40 feet underground covering an area of 90 by 120 feet. A year later, it covered an entire acre and had penetrated an additional 10 feet down; the neighbors' drinking water was in danger of contamination, and Dow agreed to settle.[26]

In Louisiana, a chemical company typically applies for separate permits depending on whether they intend to release materials into the air, surface water, or ground (e.g., injection wells.) Often firms have permits for all three release media. Permits for ground or subsurface waste disposal are obtained from the Louisiana Department of Natural Resources rather than the DEQ. Coordination between these branches of government is problematic, sometimes leading to contradictory disposal decisions that primarily favor the corporations.[27] Citizens fighting to restrict the deep well injection of chemical company waste are at a disadvantage, because the hazardous material is no longer visible. They must create collective ways to narrate the subterranean landscape in order to prevent underground waste disposal.

Beneath the surface of Louisiana's Mississippi Delta there are alternating deposits of sand, clay, and some limestone. As depth and pressure increase, the sands become sandstone and the clays become shale. From about 100 to 3,500 feet down, the sand beds contain accessible fresh water, the source of the community's drinking water. Injection wells typically expel hazardous liquid waste, under pressure, at depths ranging from 2,500 to 6,000 feet. The question is, What happens to this material once it is in the ground? The companies, without exception, use computer-generated modeling to predict underground fluid behavior.

The citizens' first narrative strategy was to challenge this approach. In 1991 BASF, a large multinational chemical firm, applied for a permit renewal to continue disposing hydrochloric acid, a by-product of its manufacturing processes, into an injection well. In an official hearing held to allow public input, chemical plant worker and community resident Kernest Lanoux said, "We are opposing the injection well . . . there's no proof that the toxic waste is not moving. All we have is computer models and these are the same models that said the Challenger Space Shuttle wouldn't blow up. We don't trust models."[28] Lanoux explained that three of his immediate neighbors, including two young mothers and his own father, have cancer; he feels its only a matter of time before he is stricken too.

The corporate models of the underground landscape describe its sedimentation as forming discrete pockets or "containers" of sand sealed off from surrounding water supplies and other flowing strata by dense, impermeable clay. The absorbent wet sand pockets are presented as perfect disposal vessels for hazardous waste, which by law must contain the waste without leaking for ten thousand years. At the BASF hearing, citizen-activist Donna Carrier disputed the reliability of this subterranean model, describing the clay barrier as inconsistent and full of millions of tiny cracks and holes.[29] While we may understand the movement of water underground, does that necessarily translate to how heavy organic chemicals travel underground? Citizens cited Brad Hanson's research for the Louisiana Geological Survey, which questioned injection wells in this delta subsoil area, because of the potential pathways for upward migration of the waste. Shortly after public release of his research, Hanson was ordered by state officials to stop his work. Lack of manpower and duplication of research elsewhere were cited as reasons for ending his investigation.[30]

Paul Brierre, a student attorney who is helping the community, asked the state to imagine a synergistic effect. What happens when the chemicals from the surrounding industry's injection wells mix with the eight-and-a-half million gallons of hydrochloric acid that BASF is putting in the ground each month? The companies predict a "cone of influence" for the waste, which is an estimation of the size and flow of the waste plume over time. Brierre compared several permit applications for injecting waste from the industries adjacent to BASF. They all used computer models to predict underground behavior and migration, yet they give contradictory predictions regarding the direction of flow. The community also raised the issue of migration of waste both upward to the water supply and outward under neighbors' property and homes. Of concern is the state's legislative silence on underground migration of these waste plumes. Neighbors believe that waste materials under their property may expose them to unwanted litigation as well as possible harm to their drinking water. The state's position is that if you can't see it and it's not in your water, don't think about it.

When citizens testify at public hearings regarding the permitting of injection wells or of toxic releases into other media, such as air or surface water, the law requires that their questions and concerns be addressed only if they are relevant to permit application. For example, in the Department of Natural Resource's response to public comments on repermitting the BASF injection well, only "technical" concerns of the public were deemed relevant and requiring a response. Technical matters were those that addressed the permeability of the retention clay or the likelihood of waste migration. To answer the public's concerns, the state referred to the technical information submitted by BASF as part of the permitting application rather than obtaining an outside opinion like that of Hanson. It appeared to many that the state officials went out of their way to discount all information that contradicted BASF's application. This did little for the citizen's confidence in the governmental agency's willingness to act on behalf of public safety in their decision-making processes. Citizen accounts of other related concerns, for example, health issues as well as cultural and wildlife factors, were never addressed in the official response, as they were deemed to be nontechnical and therefore not important. Although the community members were quite articulate in narrating the story of their toxic landscape, the state government segregated public input into technical and nontechnical, or relevant and irrelevant, effectively silencing the citizen's holistic view of their community environment.

The official story from the state government and industry is quite different from that of the local residents. Government and industry don't see the property in question as landscape at all. Their narratives systematically divide the environment into three categories: land, air, and water. The landscape to them is efficacy of production, transportation, and disposal. It is not a home; it is a capital investment. Thus their relationship to the land is fundamentally different. Whereas the residents, mostly homeowners, view this place as constitutive of who they are, the corporations' interest in the land is only instrumental: a way to increase profit margins and satisfy investors.[31]

This incongruity in goals and values makes the dialogue between the residents and industry as mediated by the state government quite frustrating since they speak of different landscapes in different languages with different intentions. Industry refuses to address things such as smell and health, as they are intangible and unprovable. Proper state-endorsed scientific evidence of these problems is hard to come by. Instead, company officials hire scientists who issue complicated reports on the hydrology and geology of the groundwater as proof of the impossibility of contamination. They talk in terms of maximum pounds of emission releases of unpronounceable chemical compounds. None of this sounds remotely familiar to the residents who call the same landscape home.

This difference in landscape epistemologies divides these interpretive communities; they are unable to communicate across their great cultural divide. The scientific discourse of emissions, land well injections, and the various other disposal technologies are not compatible with the residents' discourse on the imposition of this technology on their daily lives. Some of them attempt to engage in the scientific realm and find it discouraging.

Darlene Genova, a Girl Scout leader and resident, spent $75 of her own money to obtain an Army Corps of Engineer report after being told that her summer outing with the Girl Scouts had to be canceled because of waste contamination. When she spoke at the environmental justice hearings, she lugged a pile of papers and documents to the front of the room. She was angry because all of her attempts to ensure public input into siting the waste disposal facility in her community were for naught. The DEQ spends "hours and weeks with the applicant polluter to help them get their permitting applications legally and technically correct, and to help the polluters stay in business," she exclaimed, "but the time for citizens just isn't there." Public notices of siting decisions do not work, according to Genova, because they "can be placed anywhere in the newspaper, in the legal section, the classified, and even the social section . . . and often they don't really refer to the final site that's been selected [so] you really don't know that they've chosen the site right next to your home."[32] The process, its knowledge base, and its language are biased, say the citizens, in favor of the polluters.

Robinson has organized her citizen campaign against toxics around issues of environmental justice. She testified that "this vicious form of racism has already resulted in the loss of the precious heritage of Reveilletown, Morrisonville, and Sunrise" and "is responsible for the untold suffering of the people" in many poor and minority communities.[33] She has found what she believes to be a mediating discourse. She resides in a nearby community that is "host" to two Superfund sites. As a scientist, she has been collecting the empirical evidence claimed lacking by state and corporate officials. Community members are given user-friendly maps on which to chart suspicious occurrences and sites. Robinson goes door-to-door enlisting residents in her documentation campaign. She then records any illnesses, dead animals, funny smells, and suspicious clouds on the maps. "Community people do not need to be scientists," said Robinson, "but they do have to have some knowledge of how science operates so they can debunk a lot of the nonsense that's out there in the name of science."[34] She then uses the material in the scientific presentation she gives at public hearings as an "expert" biologist.[35] In this way she has been able to form a bridge between interpretive communities. Though the corporate players still believe her maps do not constitute proper "evidence," the state officials are proving to have a broader view of what counts as relevant in the toxic landscape.

Through mapping, environmental problems are now tied to a specific geographic location inscribing neighborhood complaints on a representational landscape. Robinson's maps mark the emergence of new landscape texts, reinforcing public readings of the environment in a language that can begin to disrupt the dominant discourse that normalizes massive pollution. In this way, the interpretive communities that form the environmental justice movement can begin to redefine what counts as knowledge. The map becomes an intertext between human and environment that speaks about the relationship between the two. This is what Haraway advocates in suggesting that the nonhuman world be understood as an agent in the production of the new knowledge. She also claims that epistemological shifts accompany successful social movements.[36] Local production of environmental knowledge arises at the level of resident action when the people involved are able to hybridize many languages, both expert and everyday, in such a way as to reinvent their environment in their own terms, in their own words. Once out of sanitized "expert" hands, the terrain of discourse about toxicity begins to shift. Michel Foucault might say interpretation (knowledge) is shifted to bend to a new will (power) and force participation in a different game.[37] This strategy of power, mapping, and the technique of knowledge, storytelling, represent a major step toward a truly democratic, public participation in the future transformation of Cancer Alley.

NOTES

1. Feminist historian Joan Scott has written extensively on the discursive nature of experience. See "The Evidence of Experience," *Critical Inquiry* 17 (Summer 1991): 773–97.

2. Marcy Darnovsky, "Overhauling the Meaning Machines: An Interview with Donna Haraway," *Socialist Review* 21, no. 2 (1991): 65–84.

3. Stanley Fish, *Doing What Comes Naturally* (Durham, N.C.: Duke University Press, 1989), 141, 15.

4. Records of the Assistant Commissioner for the State of Louisiana, Bureau of Refugees, Freedmen, and Abandoned Lands, 1865–1869, vol. 77, Register of Applications of Freedmen for Land, microfilm M 1027, roll 34, National Archives, Washington, D.C.

5. Sydney A. Marchand, *The Story of Ascension Parish* (Donaldsonville, L.A.: n. p., 1931), 117–118.

6. Beverly Wright, Pat Bryant, and Robert Bullard, "Coping with Poisons in Cancer Alley," in *Unequal Protection: Environmental Justice and Communities of Color,* ed. Robert Bullard (San Francisco: Sierra Club Books, 1994), 115.

7. I interviewed a number of commercial/industrial realtors who asked not to be identified, given the new interest in chemical expansion in the area. They explained that only the land closest to the river was valuable to the companies, because of transportation and lack of flooding. Given the arpeut system, this was usually only 10 percent of the land purchased, and thus agricultural use had many economic advantages, including tax incentives.

8. The DEQ was pressured by the federal government in 1994 to hold a series of environmental justice hearings along the south Louisiana industrial corridor. On 19 March, 16 April, and 28 May 1994, these hearings took place and were officially transcribed; they will be noted as "EJH" followed by the date and transcription page. DEQ, *Office of Community-Industry Relations,* Baton Rouge, La.

9. United Church of Christ Commission for Racial Justice, *Toxic Wastes and Race in the United States: A National Study of the Racial and Socioeconomic*

Characteristics of Communities with Hazardous Waste Sites (New York: United Church of Christ Commission for Racial Justice, 1987).

10. Other formats used for communicating environmental justice issues in the region have been community-made videos, journalistic reports, and personal interviews.

11. Excellent arguments for the landscape as text have been presented by sociologists Thomas Greider and Lorraine Garkovich in "Landscapes: The Social Construction of Nature and the Environment," *Rural Sociology* 59, no. 1 (1994): 1–24. Geographers J. Duncan and N. Duncan make a similar argument in "(Re)reading the Landscape," *Environment and Planing D: Society and Space* 6 (1988): 117–126.

12. Scott, "Evidence of Experience," 793.

13. Mart A. Stewart, *"What Nature Suffers to Groe": Life, Labor, and Landscape on the Georgia Coast, 1680–1920* (Athens: University of Georgia Press, 1996), prologue II.

14. EJH, 16 April 1994, 33; 28 May 1994, 31.

15. Conger Beasley, "Of Pollution and Poverty: Keeping Watch in Cancer Alley," *Buzzworm* 2 (July/August 1990): 39. Cited in Robert D. Bullard, *Unequal Protection* (San Francisco: Sierra Club Books, 1994), 116.

16. Leroy Alfred, interview by author, tape recording, St. Gabriel, L. A., 3 June 1997.

17. EJH, 16 April 1994, 67; 28 May 1994, 51.

18. EJH, 28 May 1994, 41–42, 29.

19. Amos Favorite, *Witness to the Future,* interview by Branda Miller, videocassette, dir. Branda Miller, The Video Project, 1996.

20. EJH, 19 March 1994, 61; 16 April 1994, 67.

21. EJH, 16 April 1994, 28–29, 36.

22. In most of my interviews with state employees and officials, the individuals spoke freely with the understanding that they would remain anonymous to protect their jobs.

23. Shintech will now build a smaller plant on Dow Chemical's campus, twenty miles up the river in Palquemine, La.

24. EJH, 28 May 1994, 50.

25. None of my informants from the community of Morrisonville would speak without anonymity, since the Dow Chemical buyout was not yet complete. The other two settlements have been completely razed. For an excellent discussion of these events, see Jim Schwab, *Deeper Shades of Green* (San Francisco: Sierra Club Books, 1994), 253–261.

26. Liz Avants, Plaquemine resident and environmental activist, interview by author, tape recording, Plaquemine, 19 May 1997.

27. For example, in 1996 the State Land Office forbade disposal of radioactive oil field waste on state lands; however, both the DEQ and the Department of Natural Resources continued to permit deep well injections of such waste on state property. For a complete story and review, see Peter Shinkle and Bob

Anderson, "Some Companies Believed Disposals Legal," *Baton Rouge Sunday Advocate,* 18 February 1996, sec. A1, 9.

28. Louisiana Department of Natural Resources, Office of Conservation, Injection and Mining Division, *Public Hearing RE: Application of BASF Corporation–Geismar Facility, November 14, 1991* (Baton Rouge, La.), 11.

29. Public Hearing, BASF–Geismar Facility, November 14, 1991, p. 29.

30. Bob Anderson, "Geologist's Well Hazard Work Halted," *Baton Rouge Morning Advocate,* 12 November 1991, sec. B, 1–2.

31. Margaret Jane Radin posits that property owned by persons is fundamentally different from property owned by nonpersons in her book *Reinterpreting Property* (Chicago: Chicago University Press, 1993). She argues that the two types of ownership should be treated differently by the legal system.

32. EJH, 16 April 1994, 11–12, 20.

33. Louisiana Department of Environmental Quality, Air and Hazardous Waste Division, *Public Hearing RE: Supplemental Fuels, Inc., August 6, 1993* (Baton Rouge, La.), 112.

34. Florence Robinson, telephone interview by author, tape recording, Alsen, La., 31 July 1997.

35. EJH, 1994 March 19, appendix.

36. Darnovsky, "Overhauling," 75.

37. Michel Foucault, Discipline and Punish (New York: Vintage, 1979) 27–28, 304–305.

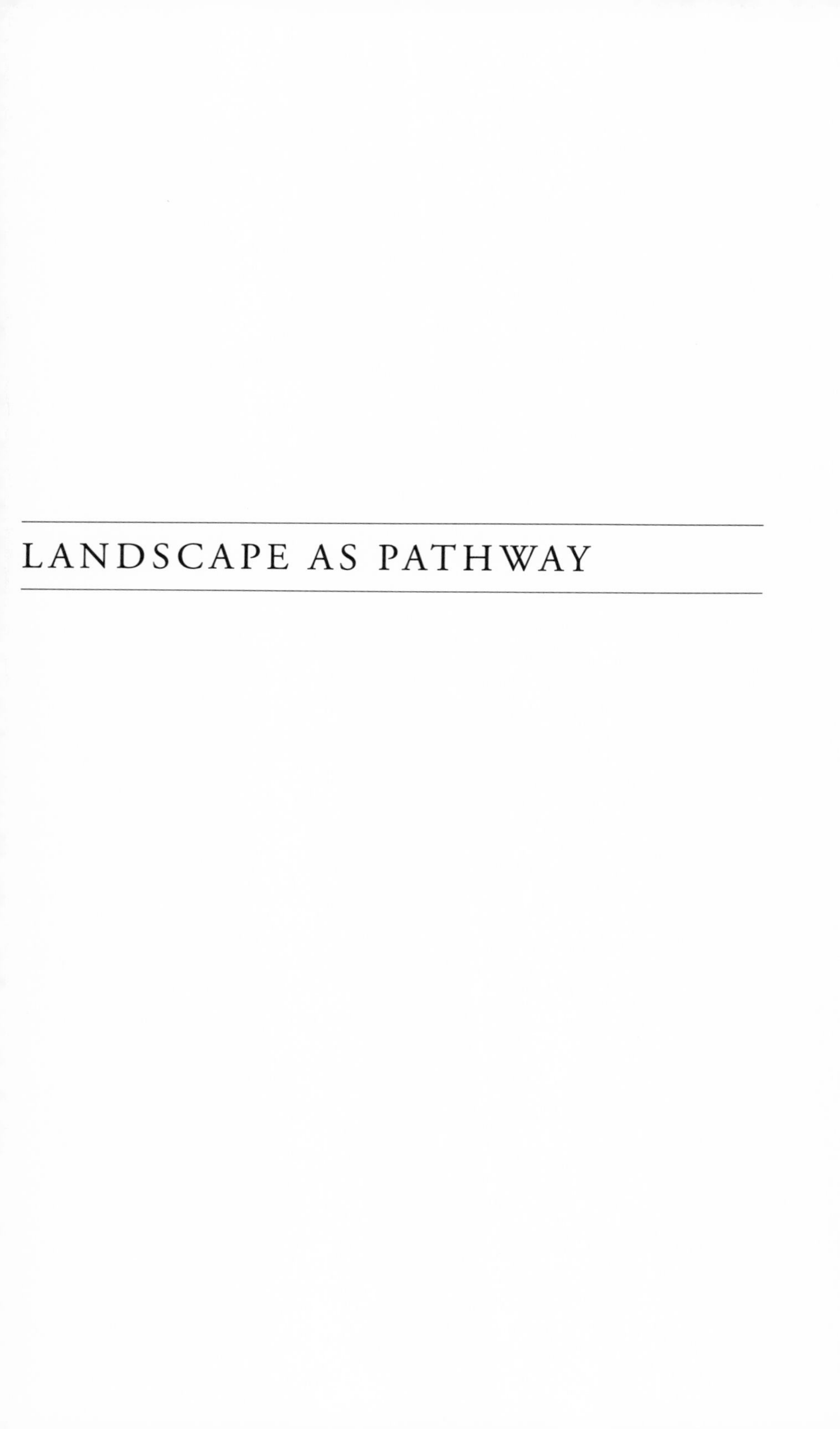

LANDSCAPE AS PATHWAY

10

Benton MacKaye's Appalachian Trail

Imagining and Engineering a Landscape

MARK LUCCARELLI

In 1921, Benton MacKaye, a professional forester, published an article in the *American Institute of Architects Journal* in which he proposed the construction of a trail through the Appalachian Mountains, the backbone of a proposed system of recreational and subsistence camps. He called it "an experiment" in getting "back to the land," a way to provide for "various kinds of non-industrial activity" and "a refuge from the scramble of every-day worldly commercial life."[1] MacKaye's proposal languished for a few years before being revived by local hiking clubs that organized as the Appalachian Trail Conference, but much of the actual construction of the trail, which covers 2,159 miles and traverses twelve states, was accomplished by the Civilian Conservation Corps in the 1930s.[2] The trail is maintained by the member hiking clubs of the Appalachian Trail Conference, and its future depends therefore on voluntary associations of wilderness enthusiasts who are now responsible not only for the trail's upkeep but for organizing the political clout necessary to preserve it.

Recently, I found on the Appalachian Trail web site the full text of MacKaye's original article, along with various accounts by outdoor recreation enthusiasts of their journeys, including journals and how-to manuals. The trail belongs now to those who seek the wilderness experience. The experience of nature they seek depends largely on the exclusion of human activity and influence. Prime wilderness can be defined as undisturbed areas. In their well-respected guide *Wilderness Areas of North America,* Ann and Myron Sutton point out that the Appalachian

Trail does not have an official "wilderness designation," but nonetheless it makes possible the experience of an extended wilderness trek, a journey made even more valuable by the fact that the hiker is simultaneously adjacent to and yet obscured from "the outposts of megalopolis." The trail is commonly understood as a path back to nonhuman nature or the wilderness. The Wilderness Act of 1964 which helped give legal status to wildlands and became a pillar of official environmentalism defines such land in the negative, as "uninterrupted and nonmanipulated natural environments," that is, as landscape without a human presence. This is nature with all the characteristics we commonly ascribe to the primeval wilderness—pristine, undisturbed, beautiful, self-contained, and self-sufficient. What is particular about the Appalachian Trail in the formulation given by the Suttons is the trek itself: "The Trail was not intended to be easy—it climbs into high rocky reaches, goes over rough terrain, plunges precipitously down steep slopes." But such hardships are well worth it, for they grant the trekker entrance into "the primitive America through which [the Trail] passes."[3]

Yet MacKaye's original plan for the Appalachian Trail differs dramatically from the use made of it today. Whereas today only a person with a taste for a strenuous outing would venture along a trail equipped with nothing more than primitive lean-tos, MacKaye actually proposed the construction of a string of "shelter camps," which he said should be "located at convenient distances so as to allow a comfortable day's walk between each" (AAT 330). The camps would contain inns equipped for sleeping and, in some instances, serving meals, after the fashion of Swiss chalets. There would have been a permanent population engaged in the business of providing services, along with other economic activities. Recreation was the economic foundation of his plan for ecological restoration and resettlement. For in addition to what we would call today "ecotourism," the trail could be the genesis of small-scale agriculture and timbering enterprises: a whole new/old economy in vital relation to its natural environs.

For MacKaye, then, the Appalachian Trail was not merely a wilderness walk or a chance to exercise outdoor skills; he actually envisioned the trail as the backbone of a plan for redevelopment—though he uses the word "development"—that would make the Appalachian backbone of eastern North America the necessary complement to the "overly commercial life" of American urban civilization. The fact that MacKaye's proposal for redevelopment derives its economic impetus in large part from ecotourism puts it in clear relation to the urban economy. This distinguished it from the kind of agrarian community development schemes that had been a common feature of proposals centering on irrigation technologies. These early proposals (beginning in the late nineteenth century) had been given impetus by passage of the Newlands Act of 1901, which mandated the construction of federal water projects; the proposals looked to re-create the economy of the early republic.[4] In contrast, the Appalachian Trail project anticipated the coming of a post-industrial economy. The trail was also to be the foundation for an alternative to existing social relations. It was to be, MacKaye writes, a "retreat from profit," an environment that is simultaneously natural and social—an opportunity to find a genuine relation to nature, including a different kind of economic relation.

MacKaye's work is unremarkable in its adherence to certain conventions of pastoralism, but he goes well beyond these conventions in positing a complex interactive relationship between the nonhuman natural world and human communities. In this respect, I find MacKaye's Appalachian Trail an interesting commentary on the attempt by some contemporary scholars of the environment to rethink the perception of nature as wilderness. Tracing the invention of wilderness as a sublime landscape, a symbol of the divine presence on earth, William Cronon points out that just as the literary celebration of natural landscapes gathered momentum in the nineteenth century, the technologically driven conquest of the natural environment culminated in the settlement and exploitation of the western lands. "Wilderness" was a creation of the culture that had viciously exploited the land; it designated land that was supposed to be free of human influence and therefore untainted by it. Cronon puts it this way: "The trouble with wilderness is that it quietly expresses and reproduces the very values its devotees seek to reject. The flight from history that is the very core of wilderness represents the false hope of an escape from responsibility, the illusion that we can somehow wipe clean the slate of our past and return to the tabula rasa that supposedly existed before we began to leave our marks on the world." In effect, by investing nature with "religious" transcendence, Cronon argues, the environmentalist project has been left vulnerable to nostalgic projection. Historically, Cronon adds, the conceptual space of "wilderness" has been fertile ground for conservative cultural motifs. Wilderness became the realm of the "rugged individualist,"[5] the pioneer-adventurer, and came to stand for a set of values that are in opposition to modernity and therefore incapable of responding to the real challenges to the environment created by human use of natural resources.

Arguing along similar lines, Carolyn Merchant proposes that we replace the predominant preservationist ethic with a "partnership ethic," defined as "a relationship between a human community and a nonhuman community in a particular place, a place that recognizes its connections to the larger world through economic and ecological exchanges . . . people would select technologies that sustained the natural environment by becoming co-workers and partners with nonhuman nature."[6] This is a vision of the coexistence and interaction of the human and nonhuman that depends on *use* as much as *nurture.* Or as Ken Olwig puts it, in the long run, the protection of natural landscapes and nature requires the caring involvement of a "productive human community."[7]

We can see MacKaye's project as a commentary on this discourse. His work shows (1) that the idea that preservation is best accomplished in relation to use is not new and (2) that, with due respect to the positive aspects of Cronon's argument, it simply is not true that holding nature in relation to the deepest insistences of self is inevitably connected to the retrogressive cultural impulses he identifies. I see MacKaye as a central figure in a tradition of American writers about nature and technology—including Frederick Law Olmsted, Lewis Mumford, and Frank Lloyd Wright—who are concerned with finding a new way of inhabiting the land. The interesting point about these writers, most true of MacKaye himself, is that they embrace the Emersonian tradition that celebrated nature as inspired. They reflect what Leo Marx calls the "pastoral impulse" in American culture—a retreat to nature in the face of an increasingly complex society; yet, as Marx reminds us, this is a retreat that is nonetheless "a serious criticism, explicit or implicit, of the established social order."[8] In short, their criticisms and constructive proposals—based on the partnership ethic—are built from the Emersonian reenchantment of the world combined with a pragmatic vision. In my view, MacKaye's work stands as testimony that an understanding of the experience of the wild is a valuable part of a valid environmentalism.

Entitled "An Appalachian Trail: A Project in Regional Planning," MacKaye's 1921 piece is half pastoral elegy and half policy proposal. He links this pastoral vision for the eastern one-third of the United States to two policy proposals: (1) the relocation of a portion of the urban population to the new recreational and productive communities to be built in the region and (2) the rebuilding of the regional economy on the basis of recreation and community-controlled agricultural and timbering enterprises. Later, MacKaye would link the concept of community control to federal ownership of the lands and federal capitalization of hydroelectric power as further steps to be taken in the redevelopment of the regional economy.[9]

Alongside these policy proposals there is an unmistakable pastoral component. The proposed Appalachian Trail would be a "skyline" that "would overlook a mighty part of the nation's activities" (AAT 326). Standing astride the trail, one would visually and metaphorically overlook the great natural areas of eastern North America—the North Woods of Maine, the Green Mountains, the Catskills, the Alleghenies, and "the Southern Appalachians where [we] find preserved much of the primal aspects of the days of Daniel Boone," including the "great Carolina hardwood belt" (AAT 327). The trail provides an aesthetic "prospect," a figurative overlook that imaginatively enfolds "the big [industrial plants] between Scranton and Pittsburgh" and the "chain of smoky bee-hive cities from Boston to Washington" into a vast natural region (AAT 327). Furthermore, the "secluded forests, pastoral lands, and water courses" that encompass the trail would serve as a retreat, "a breath of real life" for urban dwellers. At the same time, the trail-as-overlook permits us to see "flowing to waste, sometimes in terrifying floods, waters capable of generating untold hydro-electric energy and of bringing navigation to many a lower stream" (AAT 327). As MacKaye presents them, the streams seem to cry out for proper exploitation.

It is not difficult to see how a critical reader might interrogate MacKaye's work. One might argue that MacKaye gives us a version of what Merchant calls the "heroic recovery narrative." The heroic white male (in this case the regional planner and engineer) initiates the redemption of the wild, "female" nature. The language of masculine heroism is evident in MacKaye's text: "We want the strength of progress without its puniness" (AAT 325). Apparently, he wants us to exercise the virtues of scouting while we recover the memory of Daniel Boone's heroic discovery of the West. In addition, MacKaye's narrative follows the transformation of the landscape from wild to cultivated. From this perspective, MacKaye's work must be seen in the context of Western culture's relationship to nature, in which brutal exploitation was justified as salvation of nature. Reading MacKaye in this light, we might argue that the exploitation of water for the generation of hydroelectric power can be likened to the discourse that accompanied settlement of the West: they share the imperative of remaking wild nature as a garden. The hydroelectric dam is an updated plow; the potential of water power lies dormant in MacKaye's representation, much like the land awaiting the plow. In this sense, he comes dangerously close to the nineteenth-century American cultural motif that, according to David Nye, conflated preservation and transformation of the natural world.[10]

I do not discount these criticisms, but I see an opening for a different reading when looking at MacKaye's work in his own terms. Fundamental in this regard is the distinction he makes between "old" and "new explorations." The old explorers are Merchant's heroic figures. MacKaye's criticism of these explorer-heroes is tempered by his historicism. The old exploration was concerned with "actualities," that is, with finding the existing resources of the world to better exploit them. MacKaye knows far better than Perry Miller that the real purpose of the "errand into the wilderness"[11] was "the restless European energy" for trade and markets that manifested itself in "various streams of European civilization pushing towards the corners of the earth."[12] But MacKaye sees the culture of the pioneer-hero as part of material achievement of Western civilization, an achievement he feels must be acknowledged. Therefore, his position is double-sided. On the one hand, he is critical of the old exploration and posits a *new exploration* that will concern itself with natural "potentialities" and aim at developing a careful relation to the land, founded on detailed empirical and imaginative observation, and recognizing the value of natural processes in their own right. On the other hand, MacKaye is a pragmatist, certain of the need to bring new environmentalist ideals into relation to political realities while simultaneously beginning the process of redefining the technological politics that underlie those realities.

My larger point is that while there is room for disputing the theoretical implications of MacKaye's work, such a reading tends to obscure its imaginative, historical, and practical significance. Imaginatively, the work invites us to recover the Emersonian idea that human beings and our works (including our built environments) are closely related to nature. Historically, the proposal is a significant response to the wrenching changes born of an expanding capitalist economy; MacKaye's proposal for regional reconstruction was an expression of the reformist agenda of the Progressive Era. Practically, MacKaye set out to do what he could, working within the pragmatic possibilities available in a given historical era, to reimagine and reconstruct the world.

MacKaye's new exploration begins with meeting nature as different, as other, as the wild to be encountered and exchanged with, rather than as a wilderness to be conquered. The origin of his perspective was in the very conception of "wilderness" that we have heard the criticism of: MacKaye begins with the notion of nature as the "sublime" or sacred: "The primeval environment," he tells us in *The New Exploration,* "is bequeathed to us by God" (NE 58). Rather than taking this as an opportunity simply to admire a passive, abstract landscape, he sees it as an invitation to actively engage the land.[13] MacKaye quotes Aldo Leopold approvingly: "The first idea is that wilderness is a resource . . . a distinctive environment which may, if rightly used, yield certain social values" (NE 202–203). For MacKaye, then, the wild is actually an experience of "living in the open," an experience that teaches the "art" of living: "This art embraces the equivalent of all athletics and far more; it is contact not alone of man to man but of man to the whole of nature: camping is rude industry, and home-making; 'hiking' is incipient exploration; the song around the camp is the seed of folk-play and of human melody" (NE 204). MacKaye's notion of the wild incorporates even private activity—exploring, hiking, observing—into a social context.

In *The Poetics of Space,* Gaston Bachelard tells about the power to "relive dynamically" in a way that mediates between the real and the symbolic. He quotes George Sand: "What is more beautiful than a road? It is the symbol and image of an active, varied life."[14] The Appalachian Trail is that road for MacKaye: it is a symbol of nature and of a route into the interior of America, a route into the past. It is enchanted—what MacKaye means by the "primeval"—but it is also real: a direct encounter with nature as other, with a different, challenging world that requires the exertion of individual and social skills. He implies that the "discovery" of nature occurs through its "re-creation" and that valuable recreation enfolds leisure in purposeful activity.

When astride the trail, MacKaye reimagines the world of big cities and industry through an aesthetic "prospect," which is a technical term in landscape architecture for a perspective intended to give rise to contemplation. MacKaye makes wonderful use of it: the trail becomes a figurative prospect that imaginatively enfolds the industrial landscape into a vast natural region. By an act of imagination, MacKaye naturalizes the mechanical forces that gave rise to industry and the industrialized, mechanized cities. He perceives them in a larger natural context of the surrounding lands. We can see here that wishful projection—in this case, that the natural contour and force of the land are larger than the urban-industrial built environment—is not always escapist. It helps us to reimagine the world as a necessary first step in coming to terms with it. MacKaye's message is simple: seeing the city this way, we will change it, if we can find the necessary political will to create designs and economies appropriate to the ecologies of urban regions.[15]

MacKaye proposes that the Appalachian Trail become a "spearhead" of a new way of life that gains strength through its marginalities in time and space. He sees the reclaiming of the provinces as an essential opposition to an "iron civilization," a world gone mad. "The forces set loose in the jungle of our present civilization," he warns, "may prove more fierce than any beasts found in the jungle of the continents" (NE 226). These are the forces of economic growth and profit that lead to the uprooting of populations and the metropolitan migrations. Writing in the 1920s, MacKaye isn't foolhardy enough to forecast a new economic order, but he seeks to demonstrate the possibilities. He does so by formulating the Appalachian Trail as a domain marginal to metropolitan time. He imagines it as the creation of volunteers, working in their leisure time: "The customary approach to the problem of living relates to work," he says. Why not consider the problem in relation to play? "There is," he adds, "an enormous power [in] the spare time of our population" (AAT 325–326). Seizing leisure time, the trail can imaginatively re-create life. The trail is also a domain marginal to metropolitan space. It exists in the forgotten hinterland of Appalachia, one of the most neglected and brutally exploited regions in America. The trail proposal also represents MacKaye's understanding of the need to reconfigure ex-urban space in what he rightly understood would become an auto-dominated age. The trail is accessible by automobile, but not entirely so. By limiting access of wheeled vehicles—that is, by conceiving the heart of his redevelopment project as a *walking* trail—MacKaye hoped to use but control the new transportation technology. Finally, he hoped that in spurring the ecological and economic revitalization of a depressed and ecologically abused region, the trail project might stand as a model for a different conception of (social) life. Indeed, as MacKaye makes clear, altering the established "metropolitan" regime of time and space has ethical and political implications. Since the trail project was meant to be managed by communities of workers, MacKaye could hope that we might move from a regime of "competition and mutual fleecing" to one of "cooperation and mutual helpfulness" (AAT 325).

In the very first paragraph of his article MacKaye tells us, "The problem of living is at bottom an economic one" (AAT 325). There is no way to construct an alternative way of life without creating a real, that is, economic solution. Yet MacKaye was well aware of the dilemma of capitalist civilization: an increasingly interdependent economy built on greed and exploitation that has become entrenched as the engine of world economic development. "Appalachian America promises to be a strenuous battleground," he warns. "The next generation may see in this region the greatest eruption ever of iron civilization: the 'Backflows' from our metropolitan centers, big and little, may coalesce into a lava flow, or else . . . into a modern glacier whose iron fabric may do to human life and aspiration what the ice-sheet did to life in other forms" (NE 118).

MacKaye is well aware that industrial capitalism increasingly threatens the natural world. He sees technology as a threat and recognizes that the culture's appropriation of "nature" through the process of suburbanization is a real threat to both human aspiration and the nonhuman natural world. The nature of modern production and distribution has resulted in the blurring of boundaries—a process in which the urban economy overwhelms its historical borders, escapes the shell of the city, and threatens to engulf all that lies before it. Here, blurring boundaries is based on greed and unthinking exploitation. The purpose of regional planning is to control this threat of technological Armageddon. Yet MacKaye sees the threat not in technology itself but in those who controlled it: "Concentrated business, concentrated capital, concentrated power, concentrated politics" (NE 149). As we have seen, the purpose of regional planning projects, as exemplified by the Appalachian Trail, is to create spaces for the establishment of different kinds of technological, economic, and cultural regimes. For MacKaye, nurtured in the Progressive Era with its veneration of experts and its certainty about the ability of reformers to control the purposes of the centralized state, regional planning would be carried out by the federal government. It could incorporate modern technologies, such as hydroelectric dams, because the engineers would safely see to it that such technologies would be controlled and used in the public interest. The history of state-sponsored industrial development in Appalachia shows that he was mistaken in these assumptions.[16]

Nevertheless the Appalachian Trail as a project in regional planning retains its appeal. And for two reasons. First, it is built on a narrative of recovery in which human communities come to identify with the natural communities surrounding them. Here is an example of how the very process I described above—the eradication of traditional boundaries between city and country—can produce a different outcome, one based on a "partnership ethic" of "economic and ecological exchanges" between natural and human communities. Second, MacKaye's project reminds us of the value of regional land planning in engineering or reengineering our relation to nonhuman nature. For MacKaye, planning is the instrumental key for restoring the Appalachian region he sought to bring back to life through imagination to physical and social health. Reminding us that preservation and redevelopment could restore life to a forgotten and desolate region and thereby enrich all of us, MacKaye devised a powerful tool for recovering and restoring forgotten and neglected landscapes—and the economies on which the people depend. The work of regional restoration is now underway in the Great Plains, the Pacific Northwest, central Appalachia, and the Great Basin.[17] Whether these projects are part of an ecological regionalism that might realize MacKaye's hopes for restoring original relations to the land is difficult to say. Still, they are hopeful signs that MacKaye, in his embrace of pragmatic possibility, would welcome today.

NOTES

1. Benton MacKaye, "An Appalachian Trail: A Project in Regional Planning," *American Institute of Architects Journal* 9 (1921): 325–330, see 330; hereafter cited in text as "AAT."

2. See Phoebe Cutler, *The Public Landscape of the New Deal* (New Haven, Conn.: Yale University Press, 1985), 58.

3. Ann and Myron Sutton, *Wilderness Areas of North America* (New York: Funk and Wagnalls, 1974), 2, 309.

4. Donald Worster, *Rivers of Empire: Water, Aridity, and the Growth of the West* (New York: Oxford University Press, 1992), 160–185.

5. William Cronon, "The Trouble with Wilderness or Getting Back to the Wrong Nature," in *Uncommon Ground: Rethinking the Human Place in Nature,* ed. William Cronon (New York: Norton, 1996), 80.

6. Carolyn Merchant, "Reinventing Eden: Western Culture as Recovery Narrative," in Cronon, *Uncommon Ground,* 158.

7. Kenneth R. Olwig, "Reinventing Common Nature: Yosemite and Mount Rushmore—A Meandering Tale of Double Nature," in Cronon, *Uncommon Ground,* 405.

8. Leo Marx, "American Institutions and Ecological Ideals," in *The Pilot and the Passenger: Essays on Literature, Technology and Culture in the United States,* ed. Leo Marx (New York: Oxford University Press, 1988), 152.

9. See MacKaye on the Tennessee Valley Authority in "Tennessee—Seed of a National Plan," *Survey Graphic* 22 (1933): 251–254, 293–294; reprinted in *From Geography to Geotechnics,* ed. P. T. Bryant (Urbana: University of Illinois Press, 1968), 132–148.

10. David Nye, *American Technological Sublime* (Cambridge, Mass.: MIT Press, 1994).

11. Perry Miller, *Errand into the Wilderness* (Cambridge, Mass.: Belknap Press of Harvard University, 1956).

12. Benton MacKaye, *The New Exploration: A Philosophy of Regional Planning* (1928; reprint, Urbana: University of Illinois Press, 1962), 105; hereafter cited in text as "NE."

13. See John Barrel, *The Idea of Landscape and the Sense of Place* (Cambridge: Cambridge University Press, 1972).

14. Gaston Bachelard, *The Poetics of Space,* trans. Maria Jolas (1958; Boston: Beacon, 1994), 11.

15. For a contemporary discussion of this project, see David Gordon, ed., *Green Cities: Ecologically Sound Approaches to Urban Space* (Montreal: Black Rose Books, 1990).

16. On the history of MacKaye's planning efforts in conjunction with Lewis Mumford and the Regional Planning Association of America, see Mark Luccarelli, *Lewis Mumford and the Ecological Region* (New York: Guilford Press, 1995).

17. See Deborah E. And Frank J. Popper, "The Buffalo Commons: A Bioregional Vision of the Great Plains," *Landscape Architecture* 84 (1994): 140–144.

11

"The Landscape's Crown"

Landscape, Perceptions, and Modernizing Effects of the German Autobahn System, 1934 to 1941

THOMAS ZELLER

Some of the most perplexing and rewarding discussions of Nazi Germany revolve around the contention that the dictatorship contributed to the larger process of modernization. Rather than standing as a brutal aberration in the liberal modernization of the West, Nazism is increasingly analyzed in terms of modernity, that is, in its tangled relationship with promoting and realizing an individualistic consumer society amid racial fervor and mass murder. The older argument that the Nazis unwillingly helped blur rigid divisions of class, gender, and religion has been supplemented by the view that Nazi social policy quite intentionally meant to build, in some respects, a modern social state.[1]

The National Socialist regime's supposed concern for the environment also rests on the assumptions of a purposefully modern and modernizing state. According to this view, the Nazis enacted laws and instigated practices that would make them part of the vanguard of environmental protection. Present historiographical views of an environmentally aware National Socialist regime echo contemporaneous claims. The first nationwide nature protection law was passed in 1935: German propaganda claimed that even the erection of thirty-six hundred kilometers of four-lane concrete roads, the autobahn, would not harm nature. In the eyes of one German historian, this sensitivity to ecological problems signifies the modernity of Nazi Germany one more

I am indebted to Eve Duffy, Chris Fender, Karen Oslund, and Helmuth Trischler for comments on various versions of this essay.

time.[2] On the other side of the spectrum, another scholar, struck by the obvious dichotomy of killing millions of people and protecting millions of trees, was led to conclude, "It is, of course, painful to acknowledge how ecologically conscientious the most barbaric regime in modern history actually was."[3]

Were the National Socialists really that conscientious when it came to protecting the environment? This question is not as simple as it seems. Still, interpretations of an environmentally friendly Nazi regime are often based on evidence left behind by the orchestrated propaganda efforts of the regime, not on archival sources.[4] Therefore, historians overemphasize intentions and intellectual traditions while overlooking actual effects and consequences of Nazi environmental policy. To analyze the scope and meaning of these policies more fruitfully, it is helpful to examine the foundations as well as the specific implementations of ecologically oriented measures between 1933 and 1945.[5]

The example at hand, the National Socialist autobahn, has assumed mythic proportions (Figure 11.1). According to contemporary claims, these roads blended into existing sceneries and highlighted the organic and harmonic relationship of National Socialist technology to nature. In this vein, the autobahn was supposed to be unprecedented by any other work of technology in twentieth-century Germany. The underlying questions of this essay are why these categories were used in this way and why the Nazis in their modernizing efforts combined a program for road building with a public display of concern for nature. Professional groups associated with the Nazi autobahn, namely engineers and landscape architects, contested the meaning and relevance of nature for both the design of the motorways and their role in the construction process.[6] They brought forward different concepts of German landscape in acrimonious debates. Whereas the landscape architects stressed an emotional approach to understanding and reading landscapes and roadway design, the engineers promoted a landscape driving experience that was modern only at first glance. In their altercations over landscape and road design, both groups also were vying for professional self-assertion. I will show that the apparent modernity of the Nazi autobahn cannot be separated from its racial rationale.

The program to build an extensive road system was one of the first and most ambitious plans announced by the newly established Nazi regime. In fact, the autobahn and its German inauguration have left a collective memory in Germany. This memory perceives the autobahn as one of the great achievements of Nazi rule as opposed to the horrible mass murders committed by Germans during this dictatorship. Twisted as this logic may be, it is still deeply rooted in some German minds.[7] The enduring success of this myth is due largely to the massive propaganda orchestrated by the regime. Nazi propaganda stressed the largely overblown effects of road building on the labor market; created a vision of a motorized Germany with affordable cars and an extensive system of motorways; and ignited a discourse on the beauty of the autobahn, which appeared as the embodiment of the long-awaited reconciliation between nature and technology. Since then, the myth that the autobahn alleviated unemployment has often been refuted. The second propaganda vision, Germany as a country of car owners, extended an older tradition reaching back to Weimar Germany. In terms of private ownership of cars, Germany was falling behind France and Britain, not to mention the United States, whose level of motorization would only be reached in the 1960s in West Germany.[8] Hitler grasped the opportunity and promoted a vision of individual mobility for the new regime. The effort failed, and the envisioned "people's car," the *Volkswagen,* was never produced in appreciable numbers before 1945, except for military purposes.[9]

Although there turned out to be few cars to make use of them, the autobahn was still built. German historiography has shown the inconsistency of Nazi transportation policy, pointing to the erection of a highway system of 3,625 kilometers by the end of 1941.[10] What historians have called the polycratic structure of the National Socialist regime,

FIGURE 11.1 "Speed Along German Reichsautobahnen": poster by Robert Zinner, c. 1936. In time for the 1936 Olympic Games, this poster aimed at presenting a peaceful, technologically advanced Germany to the world. Motorists from abroad were invited to share a sense of speed on the new *Reichsautobahnen.* Their individual mobility would include unprecedented, spectacular vistas and experiencing the highway as a modern monument. Before crossing the bridge, drivers could stop at a rest area and wonder at the "organic" roadway. There, the pylon bears a swastika underneath a watchful eagle controlling six cars. The German tourist board commissioned the artist Robert Zinner to stylize the pile-bridge over the river Saale between Bavaria and Thuringia and the surrounding scenery. It is remarkable to see how early the German word *Autobahn* was internationalized. Courtesy of Persuasive Images, Stanford, California.

ROBERT ZINNER
SPEED ALONG
GERMAN REICHSAUTOBAHNEN

with its fragmentation of policies in different and rivalling political agencies, obviously helped to spur road building on this scale.

Very early Hitler established a bureaucratic agency responsible for the building of the autobahn, the office of the inspector general for German roads. The inspector general assumed responsibility for the Department of Transportation and the states. His new powers therefore only aggravated the lack of coordination between differing actors and interests, such as the still powerful national railway, the *Deutsche Reichsbahn;* the Department of Finance, which had a strong interest in the railway's revenues; and the interests of road-building companies.[11] The engineer Fritz Todt, a former associate of these road-building companies and a first-generation Nazi, became inspector general in June 1933. Todt was the first nonlawyer to head a German Reich agency. He had modeled himself as a Nazi road expert by laying out a master plan for the German motorways as early as October 1932.[12]

The high speed with which the autobahn was completed reflects a variety of interests and political constellations. It was partly due to the plans that had already been made during the Weimar Republic by lobbyists for road building. At the same time, the Nazi regime wanted to build the roads as quickly as possible to achieve the desired propaganda results. The dictatorship fashioned these roads into an icon of German power and economic strength and its resurgence after the calamities of the Depression. They were to become a national symbol (Figure 11.2). The propaganda of the autobahn-building phase was extensive and impossible to escape. Posters, movies, books, magazines, and even theater plays glorified the peaceful accomplishment of an economically vigorous and organizationally unbeatable Germany. Radio shows left no detail of the building process uncovered.[13]

A closer look at this propaganda shows that the mass-media onslaught functioned on different levels: workers and their families were to be impressed by bombastic radio shows and cigarette collectibles. Educated middle-class audiences, however, who might still be skeptical about the uncouth new regime were to be won over by the autobahn's success at uniting nature and technology. This reconciliation of two hitherto separate worlds had been a prominent theme of German intellectual debates in the 1920s. In a somewhat simplified version of the argument, conservative German intellectuals feared that industrialization and a market economy would threaten and eventually annihilate

Figure 11.2 Autobahn travel in Germany in the 1930s. A car passes the swastika flag between the two autobahn lanes while two motorcyclists exit. This scene is overshadowed by the foothills of the Bavarian Alps. According to the regime's own representation, the mountain view, the roadway, and occasional Nazi paraphernalia were to climax in the ultimate technological symbol of Nazi Germany, the autobahn. Courtesy of the Deutsches Museum, Munich.

the German soul so deeply embodied in its sacred and sublime nature. However, this danger of a powerfully modernizing Germany that would destroy its inner foundation was avoidable. The *Heimatschutz* movement, after failing as a merely reactionary societal force, turned toward a more open stance in the face of capitalism.[14] To solve this dilemma, Germany only had to resort to a non-Western and noncapitalist mode of production and thus a more benign attitude toward German nature. This attitude brought forward by conservative intellectuals has been labeled "reactionary modernism" by historians. The Nazis themselves were eager to call it *Deutsche Technik,* or German technology. Accordingly, a new journal by this very name praised the autobahn as the epitome of the new era.[15]

Although there was no consistent Nazi ideology of technology among the varied strands of National Socialist thought, some common features stand out.[16] Nazi propaganda radicalized an already existing discourse and undergirded it with racist lines of argument. According to Nazi publications, the main reason that capitalism had been rampant in its destruction of German nature was that inferior races such as the Jews had been allowed to play a role in the German economy. Race was the new keyword for this discourse, and the link between race and nature was essential. Only a healthy German nature could support the offspring for the rejuvenation of the Aryan race.

As the concept of *Deutsche Technik* suggests, technology was supposed to perform a key role in this process. Under "materialistic" conditions, technology had been considered an instrument of destruction. Prophets of the *Heimatschutz* movement cited countless examples of how bridges, factories, and hydroelectric plants brutally intruded on German nature. Against this background of a cold modernism, the vision of a united organic entity of nature and technology was brought forward.[17] With the advent of the organic, architects and engineers were to play much more important roles than before. Technology, if only taken out of its materialistic context and placed into the German organic economy, could now help to enhance natural surroundings. Advocates of the *Heimatschutz* movement embraced the advent of the Nazi regime, since it promised to fulfill their organic rhetoric.

Todt, as the inspector general for roads, followed this line of argument when he declared in January 1934 that the autobahn should by no means be built like railways that had been erected decades earlier. "The railway has mostly been an alien element in the landscape. A motorway, however, is and will be a street, and streets are an integral part of the landscape. German landscape is full of character. Therefore, the motorways must assume a German character."[18] In vague terms, Todt asked for the new roads not to dominate nature but to blend into existing scenery. By the same token, the roads should give drivers spectacular views of German landscapes.

To ensure that this landscape-friendly style would be upheld, Todt's office hired a cadre of landscape architects. In the summer of 1934, Todt appointed the Munich landscape architect Alwin Seifert his advisor for landscape matters. Seifert selected fifteen colleagues and became the leader of this group.[19] The administration bestowed on them the illustrious title "landscape advocates," a name that suggests a legal dispute over questions of landscape design. As Todt defined their task, the landscape advocates would act as consultants to the road-building engineers.

These engineers, however, were employees of the national railway, the *Reichsbahn,* and therefore committed to building streets with roughly the same parameters as were used for railways. Engineers had traditionally avoided curves wherever possible, situated roads on dams, and paid little or no attention to roadside plants.

Conflicts with the landscape architects who were dedicated to blending the roads into existing sceneries were inevitable. These confrontations can be understood only in the context of the hierarchical structure of Nazi road building. Because of a shortage of skilled road engineers, Todt's agency had to resort to the railway's engineers. The national railway was organized into regional headquarters with a central board in Berlin. The system of the landscape advocates was erected alongside this structure: every regional headquarters was to be counseled by one landscape architect, with Seifert in Munich serving as the liaison to Todt. On an everyday level, this structure turned out to be awkward.[20] What made the system even more confusing to the architects was that their task was ill defined. Todt had stipulated that the regional offices would consult the landscape advocates before designing their plans. More often than not, however, the road engineers did not contact the landscape architects until after the plans had been made and the machines had rolled out to start building the roads. This negligence led to a conflict between Seifert and Todt that was fought out along ideological lines, and minute technical details of road building became the subject of serious disputes about the "Germanness" of roads.

The first example was the question of curves. Seifert believed that only a sweeping autobahn would be appropriate to the values he saw incorporated in German landscapes. Roads and bridges were to sweep elegantly over the valleys, since a straight line or straight road was, according to Seifert, unnatural. "The straight line is of cosmic origin," he wrote. "It is not from this earth and is not found in nature. No living thing can move itself forward in a straight line."[21] In Seifert's eyes, only nature itself could point to the single fitting design of the road. The landscape advocates had to understand nature's laws in a way that could not be measured, only felt, and then help design the roads according to these feelings. What Seifert was effectively asking for was the elevation of his aesthetic and artistic appreciation of nature into a valid guideline for road building and the downplaying of a scientific understanding of nature. For the landscape advocates, nature was the sublime actor and humans could only subdue themselves to its will. What is more, these eternal laws would be legible to human minds only through the design proposals of the landscape advocates. In this vein of thought, technology had to be modeled according to nature's laws. If there were no straight lines in nature, then straight motorways were simply inappropriate.

As a trained engineer, Todt disputed these arguments. Interestingly enough, he used nature as his founding principle as well. In response to Seifert, Todt wrote, "The car is not a rabbit or a deer that jumps around in sweeping lines, but it is a man-made work of technology in need of an appropriate roadway. Rather, the car resembles a dragon fly or any other jumping animal that moves shorter distances in straight lines and then changes its direction at different points."[22] The nature of a man-made technology would determine its suitability. Still, Todt resorted to animal analogies and confused his own argument about the human quality of roads with references to nature. At the interstices of landscape and technology, both engineers and landscape architects employed ideas of nature to support their respective intellectual and professional status in the building of the autobahn and in Nazi society. In these efforts, the compelling normative qualities of nature were the final argument. Since Todt had the power to define what constituted an appropriate roadway, alignment of the early autobahn consisted of straight stretches sewn together with short bends. The landscape architects ridiculed this as "zigzagging" (Figure 11.3).

Only after five years of road building and the steady admonishments of the landscape architects did the sweeping autobahn slowly come into being. Compared to the aesthetic qualities so vividly described by the landscape advocates, however, the road-building administration was much more impressed with the cost-saving effect of roads, which called for fewer bridges and less costly digging. Sweeping roads followed the inherent landscape patterns and were laid out around the mountains and in the valleys instead of over the mountains and their slopes. Furthermore, the road engineers began to prefer landscape-oriented roads because they deemed them safer than straight ones. After some untrained German drivers had suffered serious accidents, it was believed that more curves would suppress drivers' boredom and thus keep them alert. One of Todt's first reactions to the increasing number of accidents on the autobahn had been to hire a dowser with a divining rod, since

Figure 11.3 The autobahn in Saxony in the 1930s. The bulk of the autobahn in Nazi Germany used design features that belied the organic rhetoric of the regime. The dam on which this stretch near Görlitz in Saxony was built is reminiscent of railway lines—the embankment's incline is less sharp, however—and the relatively sudden curve spoke against the smooth qualities of the roads vividly described in the propaganda. Roadside planting was kept to a minimum, since the road-building administration declared it too costly. In the far distance, a lone car points to the scarcity of automobiles in Nazi Germany. Still, huge resources were allocated for the autobahn. Courtesy of the Deutsches Museum, Munich.

Todt suspected underground water lines to be responsible for causing casualties. Yet, the dowser did not provide an explanation.[23]

When declaring the sweeping autobahn appropriate for German landscapes in 1940, Todt made it clear that these standards had to be applied solely to Germany. This exclusiveness was intricately linked to racist perceptions. In a speech given to his inner circle, Todt said he could not imagine that "we should make much of an effort" to preserve "remainders of natural beauty" in conquered Poland. Similar conditions would prevail in occupied Belgium, where a "relatively speedy course" should be sufficient.[24]

The landscape advocates did not fulfill their aims when it came to the question of sweeping motorways. What is more, some regional units of the road-building administration used their expertise only after the roads had already been mapped out. In an attempt to prove their usefulness, the landscape architects insisted on planting the appropriate trees and shrubs alongside the autobahn. One design feature of the conservative landscape architecture before 1933 had been a preference for indigenous plants for gardens and parks.[25] Seifert declared that on the open landscape of the autobahn native plants were absolutely necessary. No "non-German" tree or shrub was allowed to be planted, and in his view, there existed only one fitting plant for every single spot of German nature.

This notion of a specific and appropriate community of soil and plants was taken from plant sociology, as the branch of ecology that examined natural vegetation units was called in Europe.[26] Scientists ascertained that vegetation organized itself into discrete units over time. Therefore, biologists and ecologists claimed that communities of plants and soil existed on different hierarchical levels: they started on a primitive stage only to reach a final, static level. In Europe, plant sociologists were interested in the plant association. In the United States, this association was taken to an organistic extreme. The American biologist Frederick Clements stated that "the unit of vegetation, the climax formation, is an organic entity. As an organism, the formation arises, grows, matures, and dies . . . The climax formation is the adult organism, the fully developed community, of which all initial and medial stages are but stages of development."[27] Although the European and American concepts and approaches differed, ecologists on both sides of the Atlantic believed in the feature of a steady and final community of plants.

The German landscape architects associated with the autobahn did not stop searching for this adult organism. They radicalized Clements's organic worldview by emphasizing the Germanness of the fully developed plant community. Seifert demanded that the landscape advocates must not plant "foreign shrubs" such as lilac, jasmine, park roses, douglas firs, and rhododendron, since all these plants were not indigenous to Germany.[28] The landscape advocates also expected plant sociology to help them reestablish plant associations that had vanished from the face of German landscape as a result of the utilitarianism of the market economy. "It is about time to end a century of aberration in the relationship between nature and technology," declared Seifert.[29] This usually meant planting deciduous instead of coniferous trees alongside the autobahn.

Coniferous trees had been chosen to grow timber commercially, whereas the deciduous trees of the 1930s would be selected for their natural beauty and, more importantly, their appropriateness. The landscape advocates believed that through their efforts, the erection of a new transport system, the autobahn, could lead to the restoration of true German nature and the revival of original plant communities.

In their quest for the restoration of these older communities, the landscape architects were willing to resort to scientific methods. Quite in contrast to their aesthetic and artistic approach to road design and the question of curves, which bordered on self-centered willfulness, the men around Seifert embraced the scientifically generated results of plant ecologists, which led them to one appropriate vegetation association for every unit. The architects convinced Todt's road-building administration to fund the fledgling discipline of plant sociology. Starting in 1935, ecologists analyzed the existing plant associations alongside the autobahn, produced maps, and pointed to the "natural" plant community that would exist without humans' influence on earth.[30]

These newly generated results led the landscape architects to plant many more trees and shrubs than the road-building administration was willing to pay for. At the end of 1936, in one of the most turbulent phases of motorway building, inspector general Todt was upset because the landscape advocates had placed too many plants between the lanes and alongside the autobahn.[31] Impressions of landscape, wrote Todt, should be generated by the "open spaces of landscape" and sequences of these spaces, not by planting trees and shrubs. Therefore, only 10 or 20 out of every 100 kilometers of roadways should be covered with plants.[32] After summoning all the landscape advocates to his Berlin office, Todt issued new guidelines for roadside vegetation. He defined its foremost task as enhancing the driving experience on the autobahn. The drivers should be kept in an alert state by looking at varying landscape designs. Therefore, the vegetation had to alternate between sparsely and densely planted areas. According to Todt, this was more important than blending the autobahn into the surrounding landscape. Thus, the roads would become, in Todt's words, the "crown of the landscape" opened up by the autobahn.[33]

Todt's focus on the driving experience also led him to remain skeptical in the face of the radical revisionism of the landscape advocates. He met their call for the reintroduction of original German nature with pragmatic caution. He warned that it could take some time before original plant communities were reestablished and added that, therefore, the planting of autochtonous trees should not be hastened.[34]

Besides general skepticism, there was an additional reason for Todt's hesitancy. The German Reich's expenses for the autobahn had skyrocketed. Instead of spending the estimated 600,000 reichsmark per kilometer of the four-lane autobahn, the regime paid 900,000 reichsmark per kilometer. This increase was due largely to the speed of the building process. Since the administration constantly needed to lower building costs, and roadside vegetation's scope was so obvious, with planting done at the very last stage of road building, the trees and shrubs alongside the autobahn more often than not fell prey to financial constraints. Ornamental trees would not facilitate the regime's preparation for war. The Nazi regime's total expenses for the landscape friendliness of the autobahn amounted to 800 reichsmark per kilometer in 1937, including the landscape advocates' fees and the expenses for trees and shrubs. Exactly 0.08 percent of the overall costs for the autobahn were used for the landscape advocates and their tasks.[35]

Although the concept of a specifically German technology, *Deutsche Technik,* was shared by both the inspector general's office and the landscape advocates, they differed when it came to the meaning of the landscape of the autobahn. The road-building administration was inclined to understand landscape as a backdrop for the drivers' experience (Figure 11.4). The various regions of Germany were to be interpreted in a distinctively different context of speed and space. A new appropriation of nature from a speeding car could be gained only when spectacular new vistas were designed for visual consumption. Compared to the landscape advocates' insistence on subordinating roadway design to nature's eternal laws, the engineers' approach appears to be more modern at first glance. Yet, the driving experience was to be rooted in and generated for a racial conception of landscape, a restricted version of natural beauty.

The landscape advocates' efforts were stymied by their structural weakness within the road-building hierarchy. The technological means for creating landscape-friendly roads were disputed as well. The landscape advocates' arguments for a sweeping autobahn were first brushed aside and then acknowledged only after the office of the inspector gen-

Figure 11.4 The autobahn near Salzburg in the 1930s. When German road engineers planned the autobahn from Munich to the Austrian border close to Salzburg, they relied on the Alpine imagery to create a monumental vision that would reflect the highway's aesthetic qualities. This stretch of road seems to lead directly toward the Alps and their peaks. Single trees and bushes remained in their respective places as landscape architects tried to embed the highways in the landscape. Courtesy of the Deutsches Museum, Munich.

eral started to believe in the cost-effectiveness of this design. The architects insisted on the use of indigenous plants, but their radical rhetoric collided with the pragmatic resistance of the road-building administration. The overarching dilemma of economic constraints, which became clearer after the completion of each stretch of the autobahn, contributed to the fact that the landscape advocates became less important on the inspector general's agenda.

The effects of this huge infrastructure effort of the Nazi regime were mixed. On the whole, the parameters for road design followed aesthetic principles loaded with contemporaneous Nazi ideology. Despite the holistic rhetoric, the concern for nature was exclusive rather than inclusive (Figure 11.5). Landscape features such as bogs and moors were considered to be mere hindrances, and the administration supported the development of fast and efficient blasting techniques with no signs of landscape advocates protesting.[36] Visual effects were of the foremost importance, and it is important to note that almost no thoughts were wasted on the effects of the autobahn on its closest neighbors, the animals. Still, Nazi propaganda created quite successful myths, including the one of the nature-friendly autobahn.[37] What is more surprising is that some historians still take the nature friendliness of the Nazi regime for granted.

Finally, landscape history's now famous remark that nature contains an extraordinary amount of human history, though it is often unnoticed, leads to a remarkable parallel in the Nazi discourse on nature.[38] The Nazi regime appreciated the human history in nature openly as part of their ideology, and, at least rhetorically, it tried to change the course of human history by changing nature and thereby strengthening

FIGURE 11.5 The autobahn in the Palatinate in the 1930s. The harmony with nature evoked by autobahn rhetoric stands in obvious contrast to this picture. As the four-lane road cuts through the forests of the Palatinate, it exposes the geological formations of the Bunter sandstone and divides the woods sharply. Still, autobahn literature claimed that a ride on the forestry stretches of the roads could bring urbanites closer to nature and evoke feelings of belonging in sheltering and confined spaces. Courtesy of the Deutsches Museum, Munich.

the German race. This hubris underscores the truly totalitarian nature of the regime.

In this respect, it is hard to see how the landscape policy of the autobahn can be considered a boon to the modernization of German society, unless one resorts to rather arbitrary definitions of modernization. The Nazis' transportation policy as a whole has already been labeled unmodern by one historian.[39] The autobahn itself appeared as the embodiment of conflicting elements, some of which were modern and others oriented toward reestablishing a lost natural past.[40] Some of the design features of the motorways were adapted from the parkways in the United States; however, their concept was transformed into an everyday transportation route. What is more, they had to serve the notion of Germanness, that is, a racially grounded national identity. The enlistment of a group of landscape advocates answered the call of the environmentalist movements of the day, yet the advocates' efforts were drastically mitigated in the building process of the autobahn as they contended with the engineers over professional status and expert knowledge. On a deeper level, the expertise generated by the counselors was used by the regime at will; instead of a modern, participatory democratic exchange of ideas on nature and its different usages, authoritarian, behind-the-scenes decisions gained true importance. Even the possibilities of visual consumption and a seemingly democratic, individualistic appropriation of speed were inextricably bound to nurturing a *völkisch* community of Aryan subjects on the autobahn. They were to be impressed by a strong state's capacity to unveil a nationwide transportation system and its ability to create scenic impressions. Only by acknowledging the intricacies and discrepancies of this amalgam can we understand the nature of the autobahn more deeply.[41] Moreover, the landscape of the autobahn had an ironic legacy. After 1945, attempts to bring nature back into the discourse of modernization not only were morally discredited by the association with the Nazis, but were tainted by the smell of blood and soil.

NOTES

1. On the relationship between modernity and Nazism, see Zygmunt Baumann, *Modernity and the Holocaust* (Ithaca, N.Y.: Cornell University Press, 1989); Peter Fritzsche, "Nazi Modern," *Modernism/Modernity* 3 (1996): 1–21.

The assumption of an unintentional tearing down of class barriers builds on the works of Ralf Dahrendorf, *Society and Democracy in Germany* (New York: Doubleday, 1967), and David Schoenbaum, *Hitler's Social Revolution: Class and Status in Nazi Germany, 1933–1939* (New York: Doubleday, 1966). The more recent debate draws on Rainer Zitelmann and Michael Prinz, eds., *Nationalsozialismus und Modernisierung,* 2d ed. (Darmstadt, Germany: Wissenschaftliche Buchgesellschaft, 1994). On a political level, the latter book fits into the mold of recent efforts of the German Right to construct a positive, usable German identity.

2. Michael Prinz, "Die soziale Funktion moderner Elemente in der Gesellschaftspolitik des Nationalsozialismus," in *Nationalsozialismus und Modernisierung,* ed. Rainer Zitelmann and Michael Prinz (Darmstadt, Germany: Wissenschaftliche Buchgesellschaft, 1991), 297–327, 315.

3. Simon Schama, *Landscape and Memory* (New York: Knopf, 1995), 119. Schama, of course, delivers a tour d'horizon of changing cultural meanings of landscapes over time and is not interested in the debate on modernization or in the National Socialist regime specifically.

4. Exceptions are Michael Wettengel, "Staat und Naturschutz 1906–1945. Zur Geschichte der Staatlichen Stelle für Naturdenkmalpflege in Preußen und der Reichsstelle für Naturschutz," *Historische Zeitschrift* 257 (1993): 355–399, and Karl Ditt, *Raum und Volkstum. Die Kulturpolitik des Provinzialverbandes Westfalen 1923–1945* (Münster, Germany: Aschendorff, 1988). For a recent summary of the debate on "green Nazis," see Franz-Josef Brüggemeier, *Tschernobyl, 26. April 1986. Die Ökologische Herausforderung* (Munich: Deutscher Taschenbuch Verlag, 1998), 155–178. Also, see Raymond H. Dominick, *The Environmental Movement in Germany. Prophets and Pioneers, 1871–1971* (Bloomington: Indiana University Press, 1992).

5. The landscape of the autobahn has been the subject of two recent articles, both of which are based on printed sources that lead the authors to skewed interpretations: Dietmar Klenke, "Autobahnbau und Naturschutz in Deutschland. Eine Liaison von Nationalpolitik, Landschaftspflege und Motorisierungsvision bis zur ökologischen Wende der siebziger Jahre," in *Politische Zäsuren und gesellschaftlicher Wandel im 20. Jahrhundert. Regionale und vergleichende Perspektiven,* ed. Matthias Frese and Michael Prinz (Paderborn, Germany: Schöningh, 1996), 465–498, and William H. Rollins, "Whose Landscape? Technology, Fascism, and Environmentalism on the National Socialist *Autobahn,*" *Annals of the Association of American Geographers* 85 (1995): 494–520.

6. For a general overview of environmental history in Germany see Joachim Radkau, "Was ist Umweltgeschichte?" in *Umweltgeschichte. Umweltverträgliches Wirtschaften in historischer Perspektive,* ed. Werner Abelshauser (Göttingen, Germany: Vandenhoeck und Ruprecht, 1994), 11–28; and Günter Bayerl, Norman Fuchsloch, and Torsten Meyer, eds. *Umweltgeschichte: Methoden, Themen, Potentiale* (Münster, Germany: Waxmann, 1996). For a discussion of the American scene, see the essays in *Environmental History* 1, no. 1 (1996); and William Cronon, ed., *Uncommon Ground: Rethinking the Human Place in Nature* (New

York: Norton, 1996). A constructivist approach such as the one proposed by Cronon is not considered particularly revolutionary in Central Europe, as in the United States, since the idea of "wilderness" simply would not resonate in Europe. Nature conservationists in Germany, for example, celebrated and sought to protect *Kulturlandschaften*—culturally altered landscapes—as early as in the mid nineteenth century because these landscapes incorporated values of a romanticized rural life. *Naturlandschaften,* natural landscapes and the closest concept to wilderness, on the other hand, are sparse and culturally less significant. For an intellectual history of German landscape, see Gert Gröning and Ulfert Herlyn, eds., *Landschaftswahrnehmung und Landschaftserfahrung* (Münster, Germany: Lit, 1996).

7. See Joachim Radkau, *Technik in Deutschland. Vom 18. Jahrhundert bis zur Gegenwart* (Frankfurt/Main: Suhrkamp, 1989), 308, and Gudrun Brockhaus, *Schauder und Idylle: Faschismus als Erlebnisangebot* (Munich: Kunstmann, 1997).

8. Richard J. Overy, "Cars, Roads, and Economic Recovery in Germany, 1932–1938," in *War and Economy in the Third Reich,* ed. Richard J. Overy (Oxford: Oxford University Press, 1994), 68–89.

9. Heidrun Edelmann, "Der Traum vom 'Volkswagen,' " in *Geschichte der Zukunft des Verkehrs. Verkehrskonzepte von der Frühen Neuzeit bis zum 21. Jahrhundert,* ed. Hans-Liudger Dienel and Helmuth Trischler (Frankfurt: Campus, 1997), 280–288.

10. Christopher Kopper, "Modernität oder Scheinmodernität nationalsozialistischer Herrschaft. Das Beispiel der Verkehrspolitik," in *Von der Aufgabe der Freiheit. Politische Verantwortung und bürgerliche Gesellschaft im 19. und 20. Jahrhundert. Festschrift für Hans Mommsen zum 5. November 1995,* ed. Christian Jansen and Lutz Niethammer (Berlin: Akademie, 1995), 399–411.

11. Alfred Gottwaldt, *Julius Dorpmüller, die Reichsbahn und die Autobahn. Verkehrspolitik und Leben des Verkehrsministers bis 1945* (Berlin: Argon, 1995).

12. Franz W. Seidler, *Fritz Todt. Baumeister des Dritten Reiches* (Frankfurt: Ullstein, 1988).

13. For a recent analysis of the autobahn's representations, see Erhard Schütz and Eckhard Gruber, *Mythos Reichsautobahn. Bau und Inszenierung der "Straßen des Führers" 1933–1941* (Berlin: Christoph Links, 1996).

14. Lately, the *Heimatschutz* movement has been cited as an example of social reform in a bold and sometimes overstated revisionist effort. William Rollins, *A Greener Vision of Home: Cultural Politics and Environmental Reform in the German Heimatschutz Movement, 1904–1918* (Ann Arbor: University of Michigan Press, 1997). Cf. the older German literature stressing the conservative, if not reactionary, outlook of the movement and its close alliances with Nazi Germany; for example, Arne Andersen, "Heimatschutz: Die bürgerliche Naturschutzbewegung," in *Besiegte Natur. Geschichte der Umwelt im 19. und 20. Jahrhundert,* ed. Franz-Josef Brüggemeier and Thomas Rommelspacher, 2d ed. (Munich: C. H. Beck, 1989), 143–157.

15. *Deutsche Technik* was published from 1933 until 1943 by the only associa-

tion acknowledged by the National Socialist Party, the Kampfbund Deutscher Architekten und Ingenieure; see Jeffrey Herf, *Reactionary Modernism: Technology, Culture, and Politics in Weimar and the Third Reich* (Cambridge: Cambridge University Press, 1984). For a critique of Herf's position, see Michael Allen, "The Puzzle of Nazi Modernism: Modern Technology and Ideological Consensus in an SS Factory at Auschwitz," *Technology and Culture* 37 (1996): 527–571, esp. 545 ff. Allen is correct in remarking that romantic-modernist contradictions caused German engineers in the Nazi period "few sleepless nights" (546). By stressing the intricacies of the "actual world," however, this approach risks downplaying intellectual traditions and climates that helped to shape debates and actions. For example, such an analysis would have less success in explaining the touting of the autobahn as a work of technology that would reconcile these contradictions. Also, see Karl-Heinz Ludwig, *Technik und Ingenieure im Dritten Reich* (Düsseldorf: Athenäum/Droste, 1979).

16. Helmut Maier, "Nationalsozialistische Technikideologie und die Politisierung des 'Technikerstandes': Fritz Todt und die Zeitschrift 'Deutsche Technik,' " in *Technische Intelligenz und "Kulturfaktor Technik." Kulturvorstellungen von Technikern und Ingenieuren zwischen Kaiserreich und früher Bundesrepublik Deutschland,* ed. Burkhard Dietz, Michael Fessner, and Helmut Maier (Münster, Germany: Waxmann, 1996), 253–268.

17. The classic work of this vein is Paul Schultze-Naumburg's widely popular *Kulturarbeiten,* 9 vols. (Munich: Callwey, 1902–1917). Volumes 7 through 9 deal with "Die Gestaltung der Landschaft durch den Menschen," the designing of landscapes by humans.

18. Statement of Todt, 18 January, 1934, R43II/503, Federal Archives, Koblenz, Germany (hereafter "FAK"). Translations, unless otherwise stated, are by author.

19. The Nazi Party banned one prospective coworker because of his former membership in the freemasons and kept one architect under close watch for alleged ties to the Communist Party before 1933. Virtually all of the landscape architects in the group were connected to the *Heimatschutz* movement and came out of the German Youth Movement. The majority were close to or members of the anthroposophic movement of Rudolf Steiner, which promoted organic farming and other issues of life reform.

20. Evidence on this point abounds in the papers of Alwin Seifert. See, for example, Schneider to Seifert, 30 June 1934, folder 146, Alwin Seifert Papers, Technical University of Munich (hereafter "ASP"); Meyer-Jungclaussen to Seifert, 3 June 1934, folder 138, ASP; Siegloch to Seifert, 6 June 1938, folder 153, ASP.

21. Alwin Seifert, "Schlängelung?," in *Im Zeitalter des Lebendigen. Natur—Heimat—Technik,* ed. Alwin Seifert (Planegg, Germany: Langen Müller, 1942), 114–117, 114.

22. Todt to Seifert, 26 June 1935, NS 26/1188, FAK.

23. Todt to Direktion Reichsautobahnen, 6 July 1936, NS 26/1188, FAK, Seifert to Todt, 13 November 1939, NS 26/1188, FAK.

24. "Rede des Reichsministers Dr. Todt auf der Architekten-Tagung des Generalinspektors für das deutsche Straßenwesen auf der Plassenburg," 9, 31 August 1940, NS 26/1188, FAK. For this speech, see the apologetic autobiography of Alwin Seifert, *Ein Leben für die Landschaft* (Düsseldorf/Cologne: Diederichs, 1962), 58.

25. Joachim Wolschke-Bulmahn and Gert Gröning, "The Ideology of the Nature Garden. Nationalistic Trends in Garden Design in Germany during the Early 20th Century," *Journal of Garden History* 12 (1992): 73–80. Gert Gröning and Joachim Wolschke-Bulmahn, *Die Liebe zur Landschaft,* 2d ed., part I (Münster, Germany: Lit, 1995), and part III (Münster, Germany: Lit, 1997).

26. Michael G. Barbour, "Ecological Fragmentation in the Fifties," in Cronon, *Uncommon Ground,* 233–255.

27. Clements quoted in Donald Worster, *Nature's Economy. A History of Ecological Ideas* (Cambridge: Cambridge University Press, 1985), 211. Ludwig Trepl, *Geschichte der Ökologie, Vom 17. Jahrhundert bis zur Gegenwart* (Frankfurt/Main: Athenäum, 1987).

28. Alwin Seifert, "Natur und Technik im deutschen Straßenbau," in *Im Zeitalter des Lebendigen. Natur—Heimat—Technik,* ed. Alwin Seifert (Planegg, Germany: Langen Müller, 1942), 22–25.

29. Ibid., 22.

30. This tendency is in line with a growing amount of money allocated to fundamental research in biology. Ute Deichmann and Benno Müller-Hill, "Biological Research at Universities and Kaiser Wilhelm Institutes in Nazi Germany," in *Science, Technology, and National Socialism,* ed. Monika Renneberg and Mark Walker (Cambridge: Cambridge University Press, 1994), 160–183.

31. Seifert to Schwarz, 28 October 1936, 46.01/864, Federal Archives, Potsdam, Germany (hereafter "FAP").

32. "Entwurf eines Rundschreibens," 23 October 1936, 46.01/864, FAP.

33. "Vorläufige Richtlinien über die Bepflanzung der Kraftfahrbahnen auf Grund der Besprechung beim Generalinspektor vom 11.11.1936," folder 116, ASP. The original reads, "zur Krone der von ihr erschlossenen Landschaft."

34. Todt, "Runderlaß," 27 January 1940, NS 26/1188, FAK.

35. Author's calculations according to "Reichsautobahnen Direktion, Kosten der technisch nicht notwendigen Pflanzungen und Begrünungen," 16 April 1937, R 65II/14, FAK. According to this source, 808,500 reichsmark were spent for the landscaping of the autobahn until the end of 1936. This amount is dwarfed not only by the overall costs of erecting the autobahn system, but also by the 3,147 million reichsmark extra that were spent to have one thousand kilometers ready for a celebratory opening in September 1936. "Zusammenstellung der Mehrkosten der Baubeschleunigung zur Eröffnung verschiedener Zivilstrecken am 27. September 1936," 8 March 1937, R 65II/15, FAK.

36. The blasting of moors was celebrated in the movie *Der Kampf mit dem Moor* (Fighting the Moor). A celebratory detonation was included in the administration's festivities to commemorate the three-thousandth kilometer of

the autobahn. "Generalinspektor, An die Gefolgschaftsmitglieder!" 14 December 1938, R 65I/38, FAK.

37. This might be viewed as an expression of the modernity of the regime. By creating a myth of landscape friendliness, the regime combined older elements with a new amalgam. I do not mean to downplay the regime's failure in implementing these policies, but rather to point to the intellectual history that it created. In this respect, a differentiation between modernization and modernity is helpful.

38. "The idea of nature contains, though often unnoticed, an extraordinary amount of human history," is Raymond Williams's now famous quote. Cited after William Cronon, "Introduction: In Search of Nature," in Cronon, *Uncommon Ground,* 23–56, 25.

39. Kopper, "Modernität oder Scheinmodernität."

40. In this understanding, Jeffrey Herf's notion of reactionary modernism has to be extended, as Allen ("The Puzzle of Nazi Modernism") did, in order to make sense in light of the debate on modernization and the evidence used here. It is unreasonable to deny that modern elements can be found in the autobahn's example. Therefore, the question does not evolve around the existence or nonexistence of these elements. Rather, it is important to analyze how they were connected to contradictory elements within the same process. For example, the Nazis set out to modernize the transportation system by implementing the autobahn even if it was built ahead of demand, but by the same degree, they chose design features that sought to unite inconsistent means and ends.

41. For vision and vistas, see Teresa Brennan and Martin Jay, eds., *Vision in Context. Historical and Contemporary Perspectives on Sight* (New York: Routledge, 1996) and Rudy Koshar, *Germany's Transient Paths. Preservation and National Memory in the Twentieth Century* (Chapel Hill: University of North Carolina Press, 1998), 175. Klenke ("Autobahnbau") ties the landscaping of the autobahn together with nature conservation, which was practically irrelevant in the building process. His notion of a "constructive cooperation" follows older conceptions (469). Rollins ("Whose Landscape") goes as far as citing the autobahn as an example of "effective environmental reform" (513) and fails to place it in the larger context of Nazi Germany.

TOURING LANDSCAPES

12

The Improvement of Arthur Young

Agricultural Technology and the Production of Landscape in Eighteenth-Century England

STEPHEN BENDING

> From the battlements [of Chambord] we saw the environs, of which the park or forest forms three-fourths; it contains within a wall of about 20,000 arpents, and abounds with all sorts of game to a degree of profusion. Great tracts of this park are waste or under heath, &c. or at least a very imperfect cultivation: I could not help thinking, that if the king of France ever formed the idea of establishing one complete and perfect farm under the turnip culture of England, here is the place for it.

ARTHUR YOUNG'S comments on Chambord in his *Travels of the Kingdom of France* point to radically different visions of land use.[1] Young, standing on the battlements of a French royal palace, looks out over a landscape created for hunting and leisure. Freed from the constraints of English society, he challenges the value of this archetypally aristocratic designed landscape and imagines instead a landscape transformed by "the turnip culture of England," shorthand for the methods associated with Norfolk husbandry, with the agrarian capitalism of enclosure, with the benefits of new technology, crop rotation, large-scale farming, and low rents. The largely nonproductive socioaesthetic space of the park is replaced here by Young's vision of the socioeconomic utility of farming. Indeed, Young is noted for his outspoken support for agriculture as the basis of national prosperity and his rejection of the increasing claims of a commercial middle class to that role. Claiming the greatest farmer to be the greatest man, his writing would seem to insist on an alternative account of the value of landscape to that proposed by a seemingly unproductive landscape garden.[2] It is the relationship between these two different kinds of landscape that I consider in this essay.

Having said *two* different kinds of landscape, however, I now want to complicate that, because a neat division between productive and unproductive, aesthetic and nonaesthetic, land is difficult to maintain. Ann Bermingham has argued that as the agricultural landscape of enclosure gets to look more and more like formal gardens, so gardens change to look more like some kind of "untouched" nature.[3] But inevitably there is a range of ways in which the land can be seen in the eighteenth century, and binary oppositions, while attractive, cannot be the whole picture. Then, as now, people " 'looked' in highly determined ways," which were not necessarily consistent ways.[4]

One place to begin is the justifications of the landscape garden in terms of high culture's account of the liberal arts. A liberal art is predicated on the process of abstraction. It demands an ability to perceive the general and not simply the particular; it is this ability that raises the individual of taste above the vulgar. It is a distinction between the mind and the hand, between principle and mere practice. As John Barrell has argued, it is the same ability that is used to justify landowning as the basis of political representation, because it is the ability to perceive the wider interests of the state, rather than merely those of one's own profession or position, that entitles a person to citizenship. Only the leisured landowner, released from the concerns of a profession or trade, is capable of seeing society's true interests as distinct from partisan concerns.[5] In these terms, the landscape garden presented a particularly attractive means by which the landowner could justify his status. The ability both to recognize and to create the landscape garden within the terms of the liberal arts denoted a worthiness to own land and a concomitant worthiness to the political rights it bestowed. Equally, it is a landscape that sets out to mark its own difference from the productive agricultural land around it. It has further been suggested that the landscape parks of the second half of the eighteenth century became widespread because they seemed to offer a shared aesthetic space and to assert a class identity common to both the propertied middle classes and an aristocratic elite.[6] In distinguishing themselves from economic profit, landscape gardens laid claim to their polite status, and in recognizing the aesthetic value of such landscapes, visitors laid claim to membership of polite society.

Where then does farming stand? Two broad traditions appear, which align closely with this distinction between the liberal and the mechanical, the general and the particular. On the one hand, farming is recognized as the basis of national wealth, producing genuine surplus and therefore prosperity. It is a tradition largely associated, however, with gentlemen and with great landowners, a tradition that can turn to classical literary justification in the form of georgic and pastoral; to the moral claims of country house poetry; and tautologically to the fact that, historically, those who run the country own the land. It is also a tradition that relies on some sense of distance from labor and mediates profit in terms of national rather than merely personal wealth. For a gentleman farmer, personal profit can be justified as national prosperity, part of a larger vision of moral economy and national interest. It is also a conservative position that sees in farmland the continuation of an established social hierarchy.

While the smaller freehold or tenant farmer is implicitly a part of this national wealth creation, he figures more prominently in the other major tradition, that of the rude mechanic, the nongentleman: backward, self-interested, at once stupid and cunning, suspicious of progress and the need for change. Frequently proclaimed the backbone of the nation, tenant farmers may be propertied but not polite: that is, they are routinely excluded from an emerging polite society that otherwise emphasizes the common culture of the propertied classes.[7] What also emerges in the eighteenth century is a new breed of farmers, holding a position somewhat similar to that of city professionals and likewise buying their way into polite culture. One might set this up in terms of the problematic relationship between land, technology, and profit, because what agricultural capitalism and the new technology also do is allow money to move into different hands. The inherent problem for the elite landowner is that a commercial concern with profit undermines traditional social hierarchy. And, arguably, part of what makes a moral vision of landscape fall apart is an increasing emphasis not on spontaneous production as part of nature's bounty—the country house vision—but on the manmade productivity of technological progress (where technology would include not just machinery but the technology of breeding, manuring, crop rotation, etc.). It seems to open up a landscape of social meritocracy based on the rational endeavor of individuals.[8]

Equally, what many social historians have traced is a growing capitalist tradition that rejects moral economy for commercial profit. This vision of the land as a source of profit divorced from moral responsibility becomes the butt of much conservative satire.[9] Just as the vulgar nouveau riche citizen is satirized in town, so the new "gentleman" farmer is satirized in accounts of farmhouse parlors with brand-new piano fortes. If polite culture assumes the shared interests of the propertied, it also works hard to exclude those who are too close to the creation of their own wealth. While itself the expression of new money's collision with landed wealth, it nevertheless defines itself in terms of exclusions: the ungentlemanly, the nonaesthetic—in a word, the vulgar. Into that category is placed the lowly farmer, whose all too earthy and mechanical connection with the land is frequently deemed beyond the pale of polite culture and its aesthetics.[10]

It is in this context that Young's tours of Britain appear in the 1760s and 1770s. My reading of the tours is that Young is trying to find ways of moving both agriculture and himself toward the center of polite culture.[11] He does so by offering husbandry as agricultural technology and agriculture as scientific experimentation for the national good, divorced from personal profit and therefore a happy companion to gardening, architecture, painting, and popular aesthetics. What Young struggles with, however, is that the practicalities of the new farming methods are awkwardly close to the mechanical, the ungentlemanly. Certainly in his early writing he recognizes that the training of a gentleman, with its emphasis on the general and the abstract, makes for a poor farmer.[12] The vision of landscape most commonly available for the gentleman is fundamentally unsuited to the quotidian detail of agricultural production. Equally, in the gentleman's distance from labor there remains a continuing dislike for the mud and mess of these mechanical activities: George III's personal, and much publicized, farming of his own lands was praised by some but ridiculed by many others. At the other end of the spectrum, Young, from a minor gentry family with no money, was a figure on the edge of propertied gentility. The tours, in establishing a gentlemanly vision of agriculture, were to secure also Young's own gentlemanly status.

Until late in his career, Young was a vociferous supporter of the enclosure of common land, which was a key element in both the creation of many landscape gardens and the rise of an agricultural capitalism that produced a dispossessed laboring class and the new gentleman farmer.

In aligning himself with the economic interests of improvement, Young aligned himself also with the interests of both an expanding middle class in the countryside and the large-scale landownership of the political and social elite. While these landed interests indeed found themselves moving closer together in social terms, that their interests were not synonymous becomes apparent in the tensions inherent in Young's texts.

The first of the tours was published in 1768 and entitled, *A Six Weeks Tour, through the southern counties of England and Wales.*[13] Given the stated purpose of his work, the format is itself of note. Like The Reverend William Gilpin, he chooses the polite form of the gentlemanly tour, but the efficacy of this model for a work that delineates agricultural practices in terms of tables, diagrams, statistical calculations, and detailed "minutes" of farming activity is not immediately apparent. Certainly, the merging of agricultural issues with the description of houses and gardens was recognized as a problem by a number of his readers. Young's adoption of the form of polite travel—rather than the arguably more obvious vehicle of the farming manual or specialized scientific treatise (which of course he also used)—has built into it a claim for the wider view of the independent man of property. These texts dignify themselves in terms of a gentlemanly vision; indeed the very titles—*A Six Weeks Tour* and *A Six Month Tour*—somewhat misleadingly point to the leisured status of the gentleman.[14] Gilpin may concern himself with sketching picturesque scenes as he travels around Britain, but the gentleman farmer Arthur Young appears to claim equal merit for scenes of agricultural production. While for Gilpin the land beyond the garden represents a chance to find untouched "nature," Young searches it instead for that agricultural production that underwrites national greatness. He characteristically divides his description of the agricultural landscape into an account of farming practices, tables detailing the cost of provisions and labor, detailed financial calculations, and illustrations of new farming implements. Like Gilpin, he provides "sketches," but they are sketches of "the general œconomy and management" of agricultural regions that aim to produce an image of the land and its use. At Bury, for example, he "sketches" a farm in the following terms: "250 acres, 80 of them grass; 10 horses; 2 men; 2 boys; 5 labourers; 25 cows; 60 sheep." Drainage, road conditions, and soil are all carefully detailed.[15] Even in his *Farmer's Tour,* where Young eventually decides to place the nonagricultural in footnotes, he still refers to the reader uninterested in houses and gardens as "the mere farming reader," whereas he wishes to be "more general, and of course more useful."[16]

This conflict of interests is apparent throughout the tours. Gardens and houses are often described in as much detail as agricultural practices, but Young finds himself writing works that, in adopting different descriptive techniques, point to fundamentally different evaluations and assumptions about the land. Statistics and tables of economic productivity may be a fitting account of land in agricultural use, but they are implicitly recognized as inappropriate for the aesthetic land*scapes* of the great landowners. These alternative, indeed competing, evaluations of the land Young finally finds himself unable to resolve. His tours repeatedly insist on the fundamental value of agriculture as "the foundation of every other art, business or profession," and if well-farmed land represents the economic foundations of the nation's wealth, then a landscape garden that revels in its apparent lack of productivity must be nothing more than a missed opportunity.[17]

This is the stance Young adopts openly at Chambord, but I want to consider the way in which he studiously avoids such a conclusion when visiting English gardens. It has become clear in recent years the extent to which the great landscape parks were themselves economic as much as aesthetic ventures. As Tom Williamson has argued, "Parks were homes, farms and forestry enterprises as well as being pictures. Moreover, these diverse activities were not simply fitted around or hidden away from the dominant aesthetic. They lay at the very heart of the landscape park."[18] Yet in his tours Young makes almost no mention of agricultural and economic activity within the parks and gardens he describes. This disjunction was recognized long ago by John Barrell, who places Young in the context of a rural professional class. However, whereas Barrell suggests that Young's admiration for unproductive landscape is a guilty pleasure and sets about distinguishing Young from aristocratic landed culture, my aim is to suggest some of the ways in which the polite discourse of the tours propels Young toward that culture.[19]

For an author championing *practical* agriculture, Young seems strangely incapable of ignoring the aesthetic. When he reaches a house, he frequently adopts the rhetoric of plain speaking—of calling a spade a spade—and claims to be swayed not by fashionable judgments but by his own feelings. At Holkham, when considering the well-proportioned front of the house and especially its columns, he writes, "It may be said the proportion of a pillar is stated, and always the same.—I know nothing of architecture, but view these at *Holkham* and others at *Blenheim*—I never speak by rules, but by my eyes." He then continues, "Will you excuse these criticisms from one who knows nothing of architecture, but its power of pleasing the taste of individuals—As one among the many, I give you my opinion, but I wish you would pass over all these parts of my letters, till you see the objects yourself, for I cannot give you an idea of the building clear enough by description for you to see the propriety or absurdity of my remarks."[20] Young at once offers the response of the individual even as he drags himself away from such subjectivity with his reference to "propriety and absurdity."

If the language and values of connoisseurship in art and architecture are dealt a rhetorical—if ambiguous—blow in favor of an apparently unmediated taste, something rather different happens when Young steps into the garden. The problems in Young's claim not to be swayed by fashion become distinctly more apparent, and indeed the very emphasis on feeling places him firmly within the polite appreciation of landscape. At Piercefield, in south Wales, Young provides us with one of his fullest descriptions of a garden. He writes, "Mr. *Dodsley,* with his dells and his dingles, *and such expressive terms,* might make amends for the want of a *Claude Loraine* [sic]; however, such an idea as my plain language will give you, follows."[21] But his language is far from plain; certainly, it is not the language of tables and statistics we find in his accounts of agricultural land. Instead, as Young draws on the language of feeling employed by many a polite traveler, he provides neat exercises in the sublime and the beautiful.[22] He tells us that the united talents of "a *Claud,* a *Poussin,* a *Vernet,* and a *Smith,* would scarcely be able to sketch" such a scene, yet he nevertheless attempts a verbal account even as he claims it is impossible. He adopts the very language of landscape composition, of connoisseurship, and offers a composed verbal picture:

> From [the alcove] you look down perpendicularly on the river, with a finely cultivated slope on the other side. *To the right* is a prodigious steep shore of wood, winding to the castle, which appears *in full view,* and a part of the town. *On the left* appears a fine view of the river for some distance, the opposite shore of wild wood, with the rock appearing at places in rising cliffs, and *further on to the termination of the view* that way, the vast wall of rocks so often mentioned, which are here seen in

> length, and have a stupendous effect. *On the whole,* this view is striking and romantic.[23]

Despite his claim that he does "not write to make display of description," display is at the heart of his project. His repeated assertion that he cannot describe landscape adequately, and so "will just give the particulars of which it consists,"[24] is itself working within the compositional terms of painting, of parts and whole, of general and particular. He ostentatiously apologizes that he is not a liberal artist, but in doing so he claims to be liberal by understanding the rules and demonstrating that very ability to compose. What we see here is Young's concern to demonstrate that he is a gentleman and can deal competently with the aesthetic as well as the agricultural: the aesthetic becomes a demonstration of gentlemanly status. If anything, in tours apparently shouting about the benefits of agricultural technology, the aesthetic becomes of central importance.

Thus, Young's use of the language of sentiment and composition is a demonstration of socioaesthetic competence, of polite taste. It is in this context that we should also understand the repeated suggestion that scenes are beyond description. As I have mentioned, Young inevitably goes on to describe such scenes, but the admission highlights a markedly different treatment of the landscape of the garden from the landscape of agriculture. For the latter, Young confidently claims to represent the land with tables, figures, statistical information, and accounts of farming practice. Scenes "beyond description" point to something else. At Duncombe Park, Young writes of a landscape as "too elegantly picturesque to admit description: It is a birds-eye landscape; a casual glance at a little paradise, which seems as it were in another region."[25] At such moments, Young signals clearly that he is stepping from agricultural into aesthetic space; he is indeed "in another region," one where the gesture toward Eden is a gesture also toward a separate *social* space with an alternative set of values. The recognition of those alternative values accounts in large part for the ever-present nervousness about the inclusion of garden scenes in the tours. It accounts also perhaps for the noticeable lack of agricultural evaluation when Young walks into a garden.

If we return to Piercefield, we find that part of the landscape view includes a large farm. However, it remains precisely a part of the *view* in this account; there is no attempt to merge the two perspectives. Instead, the farm becomes a part of Young's picture making:

> One of the sweetest valleys ever beheld lies immediately beneath, but at such a depth, that every object is diminished, and appears in miniature. This valley consists of a complete farm, of about forty inclosures, grass, and corn-fields, intersected by hedges, with many trees; it is a peninsular almost surrounded by the river, which winds directly beneath, in a manner wonderfully romantic; and what makes the whole picture perfect, is its being entirely surrounded by vast rocks and precipices, covered thick with wood, down to the very water's edge. The whole is an amphitheatre, which seems dropt from the clouds, complete in all its beauty.[26]

It may not be possible to exclude gardens from agricultural tours, but it is possible to exclude agriculture from the garden. Even at the Leasowes, a landscape claiming to be a *ferme orné,* Young makes no mention of farming, takes no account of wages, productivity, or land costs. Despite his claims for the greatest farmer being the greatest man, Young's tours represent no great challenge to the landowning aesthetic of patrician culture, to the cultural self-representation of the landscape garden. The garden has no need to be productive; neither is it represented as a waste. Gardens are at once set within the agricultural landscape of the tours and demonstrably apart from that landscape. They reenact both the claims for shared values and the fundamental divisions within polite culture.

It is only on rare occasions that the landscape garden merges in Young's tours with the world of agriculture. Sir James Langham's estate at Crosswick, near Hull, provides one of the few examples of agricultural practice being considered in relation to the park. Young writes, "Contiguous to the park, and separated from it by a sunk fence, in full view of the house, is the noble pasture . . . in which you see above an hundred large oxen, and 400 fatting sheep; a stroke of the eye commands above two thousand pounds worth of live stock, feeding on the waving slopes of a hill most happily situated to enrich the views from the house." Here at least is the recognition of the garden taking in the agricultural landscape beyond its bounds. The landscape in fact becomes georgic, and the livestock "enrich the views" with a value that is at once economic, aesthetic, and stercoraceous. At Kedleston, in Derbyshire, Young praises Lord Scarsdale for having improved the farmland as a result of landscaping his park, and he concludes, "This is one great national advantage of the nobility and gentry improving the environs of their houses—they are excellent farmers, whether they design it or not."[27] Ironically, if this suggests a happier socioagricultural account of landscape, it comes in the *Farmer's Tour*, a text in which Young banishes the world of art to the footnote.

Finally, as the episode at Chambord suggests, Young's views on gardens and agriculture are expressed perhaps most clearly not in England but in the relative freedom of a foreign land. Traveling through Nice, he visits a number of noblemen's gardens where oranges are grown for profit. Of this he writes, "The garden, which with us is an object of pleasure, is here one of œconomy, and income, circumstances that are incompatible . . . that open apartment of a residence which we call a garden, should be free from the shackles of a contract, and the scene of pleasure, not profit."[28] This, indeed, is how Young conceptualizes the landscape garden of England. As a champion of agriculture, he might be expected to set himself against gardens in favor of the turnip culture of improvement, but he does not. Instead, while articulating an agricultural discourse much of the time, and aligning this with a particular form of landownership, when he reaches a garden he reaches also a recognizably different space. What we see in Young's farms and gardens is the shift from economic aesthetic to social aesthetic and, with it, his attempt to place both himself and his concerns within the arena of polite culture. Young's insistent removal from farm to garden and back again demonstrates his acute awareness of the polite aesthetic of the liberal arts and his recognition of the garden as a key site in which to rehearse that awareness.

The problem this raises as a social venture can again be best explained in terms of the French tour. On visiting the King's Library in Paris, Young discovers a series of glass cases containing models of the instruments of various trades, preserved for posterity. These trades include

> the potter, founder, brickmaker, chymist, &c. &c. and lately added a very large one of the English garden, most miserably imagined; but with all this not a plough, or an iota of agriculture; yet a farm might be much easier represented than the garden they have attempted, and with infinitely more use. I have no doubt but there may arise many uses, in which the preservation of instruments unaltered, may be of considerable utility;

> I think I see clearly, that such a use would result in agriculture, and if so, why not in other arts? These cases of models, however, have so much the air of children's playhouses, that I would not answer for my little girl, if I had her here, not crying for them.[29]

Young's annoyance at the lack of agricultural technology among these models is apparent, as is his clear evaluation of the usefulness of agriculture. This is something he studiously avoids mentioning in the English tours. What Young is confronting, of course, is not simply use, as he sees it, but polite culture's assimilation of the mechanical arts. This act of assimilation itself throws light on the reason for including gardens in his tours. The tours are constantly both playing off and combining gardens with his own agricultural concerns, using one form of landscape to convey the other into the polite world. This recognition of the garden as a crucial site for social advancement means that while Young's writing appears to represent an alternative culture of the land—one that is economic, practical, and unconcerned with conventional aesthetics—in fact he attempts the polite act of merging the garden's culture with the economic culture of the field. What makes the new technology attractive, and what allows the gentleman to participate, is that it is transformed from the practical into the experimental, from mud and mechanics into science and mechanism. What Young attempts to do in his tours is offer the gentleman one more polite activity; he offers agricultural technology as the pursuit of leisure.

NOTES

1. Arthur Young, *Travels, during the Years 1787, 1788, and 1789. Undertaken more particularly with a View of ascertaining the Cultivation, Wealth, Resources, and National Prosperity, of the Kingdom of France,* 2 vols. (Bury St. Edmund's, 1792), 54.

2. Arthur Young, *A Six Month Tour through the North of England . . . ,* 4 vols. (London, 1770), 1:xv.

3. Ann Bermingham, *Landscape and Ideology: The English Rustic Tradition, 1740–1860* (London: Thames and Hudson, 1987).

4. For some of the implications of this in landscape description, see Peter de Bolla, "The 'charm'd eye,' " in *Body and Text in the Eighteenth Century,* ed. Veronica Kelly and Dorothea E. Von Mücke (Stanford, Calif.: Stanford University Press, 1994), 89–111.

5. See John Barrell, "The Public Prospect and the Private View: The Politics of Taste in Eighteenth-Century Britain," in *Reading Landscape: Country–City–*

Capital, ed. Simon Pugh (Manchester, England: Manchester University Press, 1990), 19–40.

6. Notably, Tom Williamson, *Polite Landscapes: Gardens and Society in Eighteenth-Century England* (Stroud: Alan Sutton, 1995), chap. 5.

7. Of interest, however, is William Marshall's defense of the ungentlemanly farmer for his very lack of "polite" attributes. See, William Marshall, *Minutes of Agriculture, made on A Farm of 300 Acres of Various Soils, Near Croydon, Surry. To which is added, a Digest, wherein The Minutes are Systematized and Amplified; and elucidated by Drawings of New Implements, a Farm-Yard, &c. The Whole being published as A Sketch of the Actual Business of a Farm; As Hints to the Inexperienced Agriculturalist; As a check to the Present False Spirit of Farming; And as an Overture to Scientific Agriculture* (London, 1778), xiv–xv.

8. A useful parallel is Jürgen Habermas's account of the emergence of a "public sphere" in the eighteenth century. See Jürgen Habermas, *The Structural Transformation of the Public Sphere: An Inquiry into a Category of Bourgeois Society,* trans. Thomas Burger and Frederick Lawrence (1962; Cambridge: Polity Press, 1989).

9. Of course, this was far from new in the eighteenth century. See, for example, Andrew McRae, *God Speed the Plough: The Representation of Agrarian England, 1500–1660* (Cambridge: Cambridge University Press, 1996).

10. For one of the fullest accounts of eighteenth-century polite culture, see Paul Langford, *A Polite and Commercial People: England 1727–1783* (Oxford: Oxford University Press, 1992).

11. See Langford, who points to the appearance of the term "gentleman farmer," which "had come into being to describe a new breed of tenants whose wealth made them genteel, but who lacked the proprietorial standing of the squierarchy. It had exactly the ambiguity to match the pretensions of the former with the condescension of the latter." Paul Langford, *Public Life and the Propertied Englishman, 1689–1798* (Oxford: Clarendon Press, 1991), 560. Here I am at odds with Beth Fowkes Tobin's account of Young. In her desire to create of Young a "new economic man" in opposition to aristocratic masculinity, Tobin places him together with Marshall, ignores his claims to gentry status, and takes little account of the polite context in which his tours are written. See Beth Fowkes Tobin, "Arthur Young, Agriculture, and the Construction of the New Economic Man," in *History, Gender and Eighteenth-Century Literature,* ed. Beth Fowkes Tobin (Athens: University of Georgia Press, 1994), 179–197.

12. See, for example, Arthur Young, *The Farmer's Letters to the People of England: containing the Sentiments of a Practical Husbandman, on various subjects of the utmost importance . . . To which is added, Sylvae: or, Occasional Tracts on Husbandry and Rural Œconomics* (London, 1767), 284–297.

13. Arthur Young, *A Six Weeks Tour, through the southern counties of England and Wales. Describing particularly, I. The present state of agriculture and manufactures. II. The different methods of cultivating the soil* (London, 1768).

14. In fact, Young made the first of these "tours" when he was in search of

a new farm after personal and financial problems required his removal from the family estate at Bradfield.

15. Young, *Six Weeks Tour,* 54.

16. Arthur Young, *The Farmer's Tour through the East of England. Being the register of a journey through various counties of this Kingdom, to enquire into the state of agriculture, etc.* (London, 1771), xxi. One of Young's problems, in the words of Lord Somerville (president of the Board of Trade, 1798 to 1800), remained the widespread view that farmers were "not a reading class of people." Quoted by Nicholas Goddard, "Agricultural Literature and Societies," in *The Agrarian History of England and Wales,* ed. G. E. Mingay (Cambridge: Cambridge University Press, 1989), 6:366.

17. Young, *The Farmer's Letters,* 3.

18. Tom Williamson, "The Landscape Park: Economics, Art and Ideology," *Journal of Garden History* 13, nos. 1 and 2 (1993): 49–55, 49; and see J. V. Beckett, "Landownership and Estate Management," in Mingay, *The Agrarian History,* 6:545–640, esp. 568–569.

19. John Barrell, *The Idea of Landscape and the Sense of Place, 1730–1840: An Approach to the Poetry of John Clare* (Cambridge: Cambridge University Press, 1972), chap. 2.

20. Young, *Six Weeks Tour,* 6–7.

21. Ibid., 130–145.

22. For Young's use of composition in landscape, see also Barrell, *The Idea of Landscape,* 77–78.

23. Young, *Six Weeks Tour,* 139–140, emphasis added.

24. Young, *Six Month Tour,* 2:334.

25. Ibid., 87–90.

26. Young, *Six Weeks Tour,* 131.

27. *Farmer's Tour,* 1:65–66, 190–193.

28. Young, *Travels of the Kingdom of France,* 189.

29. Ibid., 107–108.

13

The Road to Industrial Heterotopia

Landscape, Technology, and George Orwell's Travelogue *The Road to Wigan Pier*

PIA MARIA AHLBÄCK

It must have been the late 1980s or early 1990s, but I cannot remember the exact year when I visited Wigan in Lancashire, naively looking for the topological, literary, anachronistic stereotype I was not to find—the dirty, smoky industrial town of the English north. My particular literary luggage was, of course, George Orwell's cross-generic travelogue *The Road to Wigan Pier,* written in 1936 and a dear problem of mine since 1984.[1] But late-twentieth-century Wigan had been rebuilt since Orwell's days, in red new brick: it was shiny, glassy, glittering, green. There was no smell of smoke in the air, no poisoned canal running through the town. Wigan had clearly nothing to do with the landscapes of the book. Or had it? At least one factory whistle could be heard.

I made my way to the pier that never really was, except as a joke.[2] Now there was even a proper pier, almost. At least one pub at the new Wigan Pier, today one of Britain's tourist attractions, turned out to be called The Orwell. On my way back, I got lost—actually I am still on that journey. Instead of taking the fastest route to the railway station, I made a detour past to get a look at a huge industrial site, once again listening to the factory whistle while reading the name "H. J. HEINZ," well known worldwide. Here I will leave Orwell and his book for a while, returning to him when it is time. This journey is a journey in two opposite directions: it goes backward from the present to the 1950s; through the forties and thirties, with a brief visit to the twenties; and from there forward to the present again.

I did not find any coalfields but a giant factory. When H. J. Heinz celebrated its centennial in 1987, the company's Kitt Green factory in Wigan, which had opened in 1959, was the biggest food-processing factory in Europe, according to a company history with the optimistic title *100 Years of Progress.* Like most company histories, this one is a pure success story, in which Heinz is presented as the occasionally threatened (by war and competing companies, for instance) but nevertheless obvious benefactor of humanity. What is more, Heinz's products and the way they are made, above all, are clearly constructed as the virtuous example of how to produce and make good food in our age. However, the story of Heinz in general, and of its Wigan factory in particular, is not just a construction of the perfect food product in terms of nutrition, accessibility, and cleanliness. It is also a story of devotion. "Kitt Green is simply stunning. Largest food processing plant in Europe and the Commonwealth, it incorporates the most advanced ideas in factory planning and processing."[3]

The descriptions of the Kitt Green industrial site—and of the opening of it, properly framed by royalty, bishops, and flags—tell us that we are dealing with something out of the ordinary, not just a big factory owned by a really big company. We are dealing with twentieth-century Western iconography: the picturing of high technology and its many-faced manifestations in the modern production plant, which becomes the epitome of our culture's virtues: speed, efficiency, energy, power, size, and cleanliness.

> [The] official opening of Kitt Green, a grand and glittering occasion, was on the 21 May 1959. The ceremony, carried out by the Lord Chancellor, Viscount Kilmuir, was attended by 450 guests, most of whom came up from London by special train. After a dedication by the Bishop of Liverpool and speeches of welcome, guests lunched and toured the plant, while the flags of Great Britain, Canada, Australia, the USA and Holland . . . fluttered in the sunshine.[4]

What is the high-tech factory, then, this twentieth-century icon, part picture and part place? What kind of place, so different and yet so familiar? In his essay "Other Spaces: The Principles of Heterotopia," Michel Foucault writes:[5]

> There also exist, and this is probably true for all cultures and all civilizations, real and effective spaces which are outlined in the very institution of society, but which constitute a sort of counterarrangement of effec-

> tively realized utopia, in which all the real arrangements, all the other real arrangements that can be found within society, are at one and the same time represented, challenged and overturned: a sort of place that lies outside all places and yet is actually localizable. In contrast to the utopias, these places which are absolutely *other* with respect to all the arrangements that they reflect and of which they speak might be described as heterotopias.

Foucault distinguishes between heterotopias of illusion and heterotopias of compensation, the latter of which "have the function of forming another space, another real space, as perfect, meticulous and well-arranged as ours is disordered, ill-conceived and in a sketchy state."[6]

The heterotopia of the modern factory is a cultural container, a place different and separate from the surrounding culture but including a concentration of all its important characteristics and values, so strange and striking in their immediate overwhelming visibility that they are not recognizable at once. The factory, odd and remarkable place, something of a theater, turns the visitor into a surprised and slightly shaken Brechtian spectator, watching and taking part in a spectacle in which the alienation effect (*Verfremdungseffekt*) is in full and constant force, so that it suddenly becomes possible to *see* culture: culture is revealed.[7] The realization is that the high-tech factory is culture pure and perfected. But how differently organized is culture in this place of power, the modern production plant, compared with the world outside.

Following Foucault's theory, such other spaces enter into relationships with the surrounding society, constituting peculiarly timeless places, or places with a time entirely their own: "The heterotopia enters fully into function when men find themselves in a sort of total breach of their traditional time."[8] How is this "effectively realized utopia" of the high-tech factory arranged? How are its relationships with the surrounding society organized? A few examples from the Kitt Green *Visitors' Information Handbook* give an idea of some of Western culture's arrangements of "effectively realized utopia":

> Beans. There are five lines which produce 8-oz., 16-oz. and 92-oz. Beans with filling speeds ranging from 50–1000 cans per minute. The Bean lines use around 1,000 tonnes of dry beans every week and we produce over 400 million cans of Beans every year. There are around 150 people employed in the Bean production area. Baked beans are sold around the world from Sweden to Hong Kong . . . The multi-variety line produces

> around 30 million cans of Pasta Meals every year. There are 2 can sizes 8-oz. and 16-oz. and up to 20 different varieties. It is our most flexible line . . . The Distribution Centre loads approximately 150 vehicles every day and ships over a million cases every week. The Distribution Centre employs in the region of 200 people. The Centre receives around 10,000 pallets of finished goods every week from the production lines.[9]

These passages seem outdatedly and oddly Soviet, but they were fashioned in 1995, within a multinational, capitalist company. That fact seems to be revealed by the textual emphasis on variety, markets, and flexibility. Nevertheless, judging by the *Visitors' Information Handbook* of the food-processing factory that still in 1995 was the largest in Europe, one Western kind of "effectively realized utopia" is definitely built on the material that dreams of earthly transcendence are made of: speed, size, energy, and efficiency. At the same time, the high-tech food processing factory is a heterotopia of compensation. It possesses all the values of our age, which we do not find, or do not find to the extent they are found in the refined factory form, in common disordered and unregulated places, such as messily earthbound kitchens.

In the case of the factory, concepts of traditional time such as morning, evening, night, and day are actually meaningless, since light and energy are always provided and production never stops.[10] The factory relates to the rest of society through its constant flow of products transported and distributed to it, and through people doing their repetitive work in the factory, leaving it and returning to it on a regular basis, defined by factory time-order. In the case of the tinned and bottled food produced at Kitt Green, it is distributed to the rest of Britain.[11] As for the people working at the Kitt Green plant, they come from the local community, Wigan.

Today, Heinz is the biggest employer in the Wigan area, the base of British tinned-food production.[12] It was not always so. The history of Heinz in Britain stretches further back than the 1950s. Its Wigan production plant, the planning of which involved "three land owners, five farmers and the National Coal Board," has its precursor in a wartime munitions factory in Standish on the outskirts of Wigan in the 1940s. This factory the company took over in 1946, acting against the background of at least two decades of economic discourse over lost export markets, depression, and unemployment. New export markets had long been desperately needed. One example of this discourse in its postwar variant was the slogan "Export or Die!," coined by the British government. Part of the economic discourse after 1945 emphasized so-called development areas that had been particularly badly affected by unemployment. One was Lancashire, the heart of the Industrial Revolution, for too long relying on the old export industries of coal, cotton, iron, and steel.[13]

The American Heinz Corporation was firmly established in Britain well before 1946, however, having opened its first factory outside Wigan exactly ten years after Orwell's journey north. Its first modern factory had been started in Harlesden outside London as early as 1925: "Harlesden became a showplace, the last word in food processing plant and equipment, complete with canteen, flower borders, sports and recreational facilities."[14]

An "effectively realized utopia" for the 1920s, perhaps, but not so after 1959 and the opening of the largest food factory in Europe, judging by the reaction of the company historian, looking back from the late 1980s: "The 125 employees produced fewer than 10,000 tons in their first year. A fine achievement, nevertheless, when so much depended on hand work, such as precision-filling of mixed pickles and stuffed olives."[15]

Entering the British market in the 1920s, H. J. Heinz Ltd. was an early and important representative of the many new consumer industries that would take over from the old staple industries in the 1930s. These new industries would become essential for Britain's economic recovery and the later building of the welfare state, according to the social historians John Stevenson and Chris Cook, among others.[16] The thirties were undisputedly an age of prolonged depression in coal, cotton, iron, steel, and shipbuilding, but at the same time nothing less than a new industrial structure established itself during that contradictory decade.

The thirties thus formed a historical space of battling development trends and discourses. There was a lot of talk of unemployment, depression, and war; later, the thirties were referred to as the "wasted years," the "devil's decade," and the "low dishonest decade." However, running parallel to the discourse of depression, defeatism, and despair was a new optimistic spirit: this other discourse circulated ideas of "progress," "planning," and "social policy."[17]

The British thirties can be viewed not only as a historical space inhabited by different development trends and discourses, but as a historical space in which many separate and specific industrial places were formed and juxtaposed. There was the regional juxtaposition between the South of new industries and relative prosperity and the North of the "ailing giants,"[18] the staple industries, and of unemployment and poverty. The historian Charles L. Mowat writes:

> In this paradox lay the tragedy of industrial Britain between the wars. The staple industries, which were now contracting, occupied their historic and well-defined areas—the areas of the older coalfields. South Wales, parts of the Midlands, Lancashire and the West Riding of Yorkshire, the Tees-Tyne area of Durham and Northumberland, the isolated industrial area of west Cumberland, and the industrial belt of Scotland. Here they flourished or dwindled together: coal-mining, iron and steel, engineering, shipbuilding, textiles. If the newer and expanding industries had established themselves here, all would have been well.[19]

But the picture is not that simple, if we consider the particular local case of Heinz, Wigan, and district. The old-fashioned Lancashire coal mine, together with the Lancashire cotton mill, both with a history of hundreds of years as industrial places—as British industrial places par excellence—finally had to give way to quite new industrial sites: the high-tech factories of the new consumer industries. Conservative notions of industrial production had to be abandoned. Following the economic historian E. J. Hobsbawm, these notions must have been particularly strong in Lancashire, with its long and proud industrial tradition.[20] This area had once been first, best, and biggest in world industry; it had been the world's very workshop. The stereotypical industrial place, the blackened mill and mining district, was finally to be replaced by the modern industrial stereotype, the highly automated factory.

Heinz did not turn up in Wigan until 1946, except, we can presume, in the form of products on the shop shelves and garbage in the bin. But it is difficult not to recognize the interaction of separate phenomena and trends building from the mid-1920s on. When Orwell went to Wigan in 1936, the unemployment rate within the old staple trades had already reached its zenith and had begun to go down on the national level, but throughout the 1930s unemployment remained high in the poverty pockets of the northwest and northeast of England.[21] Wigan had come to a standstill, a kind of industrial interregnum, from which it was to recover only much later. There probably was a temporary recovery due to the war, but we can presume that it was not until mining was finally replaced by a completely new, different, untraditional industry that employment improved more substantially. In 1936, Wigan was waiting for its postwar multinational employer, its new industrial places, and its new industrial images. It was this waiting Orwell witnessed, arriving in the winter of 1936, during that long phase of major transition. Exactly for what Wigan was waiting Orwell did not know. Or did he?

Orwell went down a mine twice while in the north, but did he ever in his life actually enter a modern factory? We do not know for certain.[22] It seems unlikely, though, that a journalist, so interested in society, visiting and observing so many trivial places, would not have entered one. On the other hand, it is just as strange that Orwell as journalist and writer, did not mention such a thing, since most of his other visits resulted in reports of some kind. What is certain, however, is that he wrote a great deal about technology and factories, especially in *The Road to Wigan Pier.*

The historical space Orwell was talking in when he wrote *The Road to Wigan Pier* was the 1930s, with their contradictions. The decade's two discourses are reflected in his book. He is part of the depression and despair discourse, defensively conservative in his attitude toward industrial production. Orwell is out on a mission, we must not forget, given him by the left-wing publisher Victor Gollancz, and he sides strongly with the traditionalist trade union in the area he visits. In many respects he does not succeed as sociologist, his supposed role.[23] He is trapped in his own story, in his autobiography,[24] which he constructs in such a way as to give voice to overly romantic idealizations of manual labor and the working class,[25] and to explicit, rather extreme prejudices against various phenomena and groups of people.[26] But out on his personal mission, ideologically idiosyncratic and nonconformist, he also sketches a different work, one more work in addition to that of slack sociology and sentimental autobiography. We can sense Orwell's future dystopias partly developing out of his journey between heterotopias.

If we accept industrial pollution and degradation of the natural environment as the cultural expression we call text, and if we accept it as the particular textual variety we call scribble, we can say that it was in the traditional industrial centers of Europe—Lancashire and the Ruhr region in particular—that Western civilization really learned to write. The writing of the text of industrialism meant the construction of real, extensive, but still limited places, containing deep cavities in the ground; blackened ground; slag heaps; multicolored, sluggish rivers; no vegetation; chimneys; blackened buildings; and poor people, working and living there. Although scribbled, these places were definitely "arranged," in Foucault's words: this is underscored by the fact that they were culturally negotiated, by being either subject to deliberate silence or slowly circulated within social, legal, and economic discourses. Consequently, laws were passed (but often rejected first) as to the always economic but gradually also more than economic organization of these places. "Arrangement" finally meant complete cultural incorporation through the language of law.

In the German Ruhr region, the transformation of the natural environment to dirty words took place between 1850 and 1950. "The Ruhr," states the environmental historian F.-J. Bruggemeier, "emerged as an area where industry was protected, not nature, or the environment." The seventy-kilometer-long river Emscher, for instance, streams and all, was filled with concrete and effectively killed and turned into a sewage canal. This case is unique in the world. In the 1920s, industrialists wanted to define the river Wupper, south of the Ruhr area, as an "industrial river," thereby following the case of the Emscher and finalizing the violent cultural acquisition—the writing of these rivers—that had begun long before. The air in the Ruhr area was polluted to the extent that hardly anything could grow there until 1923, when industrial production stopped completely during the French occupation, and the whole area started to bloom.[27]

Lancashire, of course, was the very masterpage of Western industrialism's alphabetization project, and part of the black ABCs of northern England, in which, writes the environmental historian Carlos Flick, "grim humor held that generations of people . . . had come to believe that in nature the sky was gray and vegetation was black." Foreign competition in the late nineteenth century speeded up pollution of the air and the earth in this area, and in the 1920s, coal-burning electric generating plants entered the industrial scene, making the pollution problem even worse.[28]

Before finally coming back to Orwell and *The Road to Wigan Pier,* I must return to Foucault's heterotopology. As already noted, Foucault distinguishes between heterotopias of compensation and heterotopias of illusion. Among individual heterotopias that Foucault suggests are places such as the garden and the holiday village, but he also defines a group of heterotopias including the rest home, the psychiatric clinic, the old people's home, and the prison as heterotopias of deviance, substituting older heterotopias of crisis. If we consider that Foucault in his general definition of heterotopia claims that it is an "effectively realized utopia," the inclusion of spaces such as psychiatric clinics and prisons might seem odd. It becomes clear, though, that these places really must be what Foucault calls "effectively realized utopias," when we recognize the criterion of identification he chose, that of pleasure as social norm, which this kind of heterotopia subverts. In connection to his list of heterotopias of deviance, Foucault claims, "In a society like our own, *where pleasure is the rule,* the inactivity of old age constitutes not only a crisis but a deviation."[29] From the point of view of pleasure-seeking society, a place where the deviant can be put away consequently becomes an "effectively realized utopia." However, what Foucault calls an "effectively realized utopia" can very well equal the imagined place we call *dystopia,* but hardly the imagined place called *utopia.* Or who would suggest a prison or psychiatric clinic as the protoplace for peace, love, and justice? Applying Foucault's theory to industrial heterotopias rooted in the previous century, we can say that they are horribly dystopic "effectively realized utopias." These are the heterotopias of illusion that tell us that everything outside them is unreal. They do form an ironic "counter" to all the other real arrangements in society, they have the power of both representing and challenging these arrangements, and they juxtapose different places and localizations within themselves. The irony of the industrial heterotopia, the *Verfremdungseffekt* at work when the heterotopia opens itself up, making culture visible, reveals that this is the way our culture produces pleasure, that is, convenience and comfort. All the comfortable places outside, all culture's convenient arrangements, are thus made to seem illusory.

Experiencing the odd seemed to have been something of an existential project for Orwell, the visitor of heterotopias of deviance, prisons and common lodging houses among them.[30] "I find that anything outrageously strange generally ends by fascinating me even when I abominate it," he confesses in *The Road to Wigan Pier.*[31] In the following oft-quoted passage, Orwell finds the otherness of the industrial space he is traveling through more real than the so-called real world outside it.

> The train bore me away, through the monstrous scenery of slag-heaps, chimneys, piled scrap-iron, foul canals, paths of cindery mud criss-crossed by the prints of clogs. This was March, but the weather had been horribly cold and everywhere there were mounds of blackened snow. As we moved slowly through the outskirts of the town we passed row after row of little grey slum houses running at right angles to the embankment. At the back of one of the houses a young woman was kneeling on the stones, poking a stick up the leaden waste-pipe which ran from the sink inside and which I suppose was blocked . . . But quite soon the train drew away into open country, and that seemed strange, almost unnatural, as though the open country had been a kind of park; for in the industrial areas one always feels that the smoke and filth must go on for ever and that no part of the earth's surface can escape them . . . Slag-heaps and chimneys seem a more normal, probable landscape than grass and trees . . . For quite a long time . . . the train was rolling through open country before the villa-civilisation began to close in upon us again, and then the outer slums, and then the slag-heaps, belching chimneys, blast furnaces, canals and gasometers of another industrial town.[32]

As we have seen, the British industrial places of the first decades of the twentieth century were part of either the previous century or the future. The landscape where Orwell was moving around in during the winter of 1936 was historically complex, burdened by the heavy cultural inscriptions of hundreds of years of industrialization. It was not actually one landscape. In *The Road to Wigan Pier,* Orwell constructs landscape in two ways. On the one hand, it is one enormous other space, a giant industrial heterotopia connected to and containing not just all of British society and culture, but all of Western civilization. This is the master heterotopia of the Western world, through which Orwell travels by train. At the same time, landscape in *The Road to Wigan Pier* is many little landscapes, less generalized and less mythologized, all of them, however, industrial heterotopias of a similar kind. When Orwell enters them, their quality as entirely other, localizable but outside the culture they contain, is what he writes: "On the outskirts of the mining towns, there are frightful landscapes, where your horizon is ringed completely round by jagged grey mountains, and underfoot is mud and ashes and overhead the steel cables where tubs of dirt travel slowly across miles of country . . . One [slag-heap] in the slums of Wigan, used as a playground, looks like a choppy sea suddenly frozen; 'the flock mattress,' it is called locally."[33]

The following is Orwell's description of a landscape in Sheffield:

> A frightful patch of waste ground . . . trampled bare of grass and littered with newspapers and old sauce-pans. To the right an isolated row of gaunt four-roomed houses, dark red, blackened by smoke. To the left an interminable vista of factory chimneys, chimney beyond chimney, fading away into a dim blackish haze. Behind me a railway embankment made of the slag from furnaces. In front, across the patch of waste ground, a cubical building of red and yellow brick, with the sign "Thomas Grocock, Haulage Contractor."[34]

What Orwell writes in *The Road to Wigan Pier* is the dystopic story of an effectively realized industrial utopia of the previous centuries: the economically exploited and environmentally degraded mill and mining district. The fact that the historical space of the 1930s was shared by juxtaposed industrial places and images makes it possible for him to follow one story up with another, however fragmentary. Thus we can trace in Orwell's book the writing of the most important industrial heterotopia of this century, the ultramodern high-tech factory.

"You have only to look about you at this moment to realise with what sinister speed the machine is getting us into its power," says Orwell in his Wigan discussion of modern technology in 1936. It is an anachronistic absurdity, he claims, to believe that the future "citizen of Utopia" would "deliberately revert to a more primitive way of life and solace his creative instincts" when he comes home from his "daily two hours of turning a handle in the tomato-canning factory." And, notes Orwell, having passed the new factories outside London: the typical postwar factory is not a gaunt barrack or an awful chaos of blackness and belching chimneys; it is a glittering white structure of concrete, glass and steel, surrounded by green lawns and beds of tulips.[35]

What is this but the Heinz Harlesden branch, transported into and rebuilt in *The Road to Wigan Pier.* Out of this industrial image grows the first faint shape of a future factory, writing itself into the landscape of Wigan: the story of the most formidable "tomato-canning" factory in Europe has begun.

NOTES

1. George Orwell, *The Road to Wigan Pier* (1937; reprint, Harmondsworth, England: Penguin Books, 1989).

2. It is said that the Wigonians used to joke about the "pier," a wharf on the Liverpool-Leeds canal once used for loading and unloading coal.

3. John Aspery, ed., *100 Years of Progress* (Hayes, England: H. J. Heinz Company Limited, 1987), 41.

4. Ibid.

5. *Heterotopia:* derived from Greek *topos* 'place,' and *heteros* 'other, different.' Michel Foucault's essay on heterotopias, "Des Espaces Autres," was based on a lecture he gave in 1967 in which he introduced the idea of heterotopias. The essay was first published in the French journal *Architecture-Mouvement-Continuité* in 1984. There are two different English versions of the essay. The one usually referred to is called "Of Other Spaces," translated into English by Jay Miskowiec and published in *Diacritics* in 1984. The version I use here is "Other Spaces: The Principles of Heterotopia," *Lotus International* 48/49 (1985–86): 9–17. I am much indebted to Leena Kore-Schröder, University of Aberdeen, who introduced me to Foucault's heterotopology through her paper "John Betjeman's English Heterotopia," presented at the conference "The Revision of England in Film and Literature, 1945–1995," Kingstone University, Kingstone-upon-Thames, England, 1995.

6. Foucault, "Other Spaces," 12, 17.

7. See, for instance, John Willett, ed., *Brecht on Theatre* (London: Methuen, 1964), 71.

8. Foucault, "Other Spaces," 15.

9. *Kitt Green Visitors' Information Handbook* (Wigan, England: H. J. Heinz Company Limited, 1995), 5, 7.

10. "Shifts" are arbitrarily, that is, linguistically, connected with night and day, since work is regulated not by darkness and light but by whatever is necessary for effective output.

11. *Kitt Green Visitors' Information Handbook,* 2.

12. *Sunday Times Magazine* (London), 13 February 1983, 23.

13. *100 Years of Progress,* 29, 38.

14. Ibid., 13.

15. Ibid.

16. John Stevenson and Chris Cook, *The Slump: Society and Politics during the Depression* (London: Quartet Books, 1979), 5.

17. Stevenson and Cook, *The Slump,* 1–30.

18. Charles Loch Mowat, *Britain between the Wars, 1918–1940* (1955; reprint, Cambridge: Cambridge University Press, 1987), 275.

19. Ibid. 274.

20. E. J. Hobsbawm, *Industry and Empire* (1968; reprint, Harmondsworth, England: Penguin Books, 1984), 13–22, 172–194.

21. Stevenson and Cook, *The Slump*, 286–287.

22. It is likely that *The Complete Works of George Orwell,* edited by Peter Davison and yet to appear, will inform us in closer detail on this point. The *Complete Works* will substitute *The Collected Essays, Journalism and Letters of George Orwell,* edited by Ian Angus and Sonia Orwell (Orwell's second wife), (London: Secker and Warburg, 1968). Many of Orwell's letters were left out of that four-volume work.

23. Orwell was commissioned by Victor Gollancz to write a book on poverty and unemployment in the north of England. *The Road to Wigan Pier* was published as a Left Book Club edition in 1937. Gollancz found it necessary to write a preface in which he distanced himself from the second part of *The Road to Wigan Pier,* in which Orwell discusses class issues and socialism in provocative terms.

24. See, for instance, Anders Iversen "Orwell's Sentimental Journey from Burma to Wigan" in *George Orwell and 1984: Six Essays,* ed. Michael Skovmand (Århus, Denmark: Seklos, 1985), 7–27.

25. Part I of *The Road to Wigan Pier* ends with Orwell's romanticized vision of a working-class home. As many critics have pointed out, this scene has a nineteenth-century, Dickensian character. Dickens was one of Orwell's literary interests and the topic of one of his major essays "Charles Dickens," published in 1940.

26. Orwell's catalogues of phenomena he disliked—feminists, nudists, pacifists, and Quakers among them—have become notorious. For a feminist critique, see Daphne Patai, *The Orwell Mystique: A Study in Male Ideology* (Amherst: University of Massachusetts Press, 1984).

27. F.-J. Bruggemeier, "The Ruhr Basin 1850–1980: A Case of Large-Scale Industrial Pollution," in *The Silent Countdown: Essays in European Environmental History,* ed. P. Brimblecombe and C. Pfister (Berlin: Springer Verlag, 1990), 210–27.

28. Carlos Flick, "The Movement for Smoke Abatement in 19th-Century Britain," *Technology and Culture* 21 (1980): 29–50, 29, 30.

29. Foucault, "Other Spaces," 15, 16, 13, emphasis added.

30. In 1932 Orwell wrote three short essays called "The Spike," "Clink," and "Common Lodging Houses," all of them published in Ian Angus and Sonia Orwell, eds., *The Collected Essays, Journalism and Letters of George Orwell,* vol. 1 (London: Secker and Warburg, 1968).

31. Orwell, *The Road to Wigan Pier,* 100–101.

32. Ibid., 14–17. Generally it is the young woman critics have paid attention to and the strictly physical landscape has been less noted. The scope of this essay does not allow for it, but the passage would require a more penetrating discussion of the woman and her surroundings *together,* since she and the industrial landscape are interrelated.

33. Ibid., 97.

34. Ibid., 98–99.

35. Ibid., 189, 185, 100.

14

Recycled Landscapes

Mining's Legacies in the Mesabi Iron Range

PETER GOIN AND ELIZABETH RAYMOND

> Usually pocketed in the mountains, the mine, the furnace, and the forge have remained a little off the track of civilization: isolation and monotony add to the defects of the activities themselves . . . taking mining regions as a whole, they are the very images of backwardness, isolation, raw animosities and lethal struggles . . . from the modern iron mines of Minnesota to the ancient silver mines of Greece, barbarism colors the entire picture.[1]

Lewis Mumford's 1934 diatribe against mining in *Technics and Civilization* encapsulates a familiar view of the industry and the landscapes it creates as blasted and godforsaken, sinister and perhaps vaguely immoral in character. Our consideration of one such mining landscape, the Mesabi Iron Range in northeastern Minnesota (Figure 14.1), is part of a collaborative project in which we—one of us a historian and the other a landscape photographer—are examining both visual and cultural legacies of mining landscapes in the United States. Mindful of Mumford's traditional critique, and of its modern environmental variants, we examine the history of attitudes toward these "waste places," where technology has transformed the landscape to such an extent that, as one mining engineer describes it, "man has now become a geologic force in himself."[2] In this article, as in the larger work, we give equal weight to visual and textual evidence.

The area now known as the Mesabi Range initially attracted Euro-American occupants when its timber began to be exploited commercially during the 1880s. While iron ore was known to exist elsewhere in

Mesabi Range
Minnesota, U.S.A.

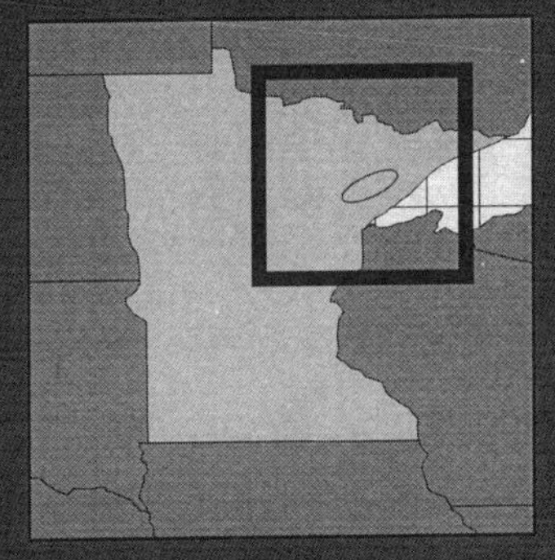

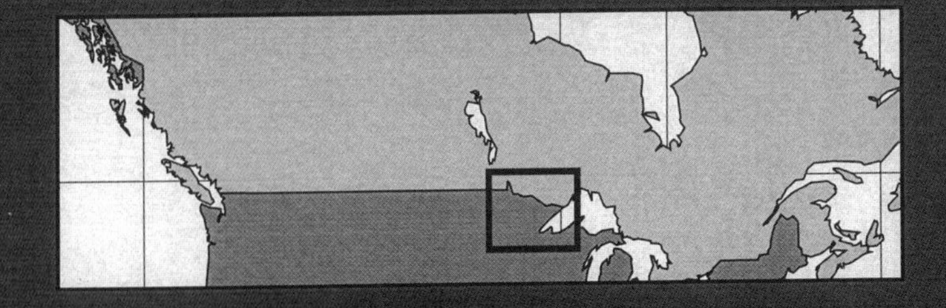

northern Minnesota, the Mesabi Range was not discovered until 1890, when the long-term prospecting efforts of Lewis Merritt and his eight sons finally paid off.[3] The Merritt brothers literally uncovered the ore when the wheels of their heavily loaded wagon cut into the soil to expose a rich, red hematite that was 67 percent iron and so soft that it could be shoveled.[4] The Merritts had spent years in search of the deposit they believed to lie in the vicinity. By virtue of this spectacular discovery they briefly became millionaires, but eventually lost all of their holdings on the range.

The deposit they had discovered, however, fully justified their years of searching. It was the largest area of the purest iron ore then known, one of the largest deposits in the world. Ore stretched for a distance of 115 miles, ranging from 2 to 10 miles in width and up to 600 feet in depth. After an initial struggle by metallurgists to find ways of using the soft Mesabi ore in their blast furnaces, it ultimately proved to be the making of the U.S. steel industry. As one contemporary railroad company prospectus crowed, "We need not worry about the market for Missabe [sic] ores, for this is the age of steel."[5]

The Merritts' discovery set off an iron boom, as miners and railroad companies rushed to gain access to the huge deposits. Initially, some underground mining was pursued, but most mining on the Mesabi was by the open-pit method. After the overburden of soil and gravel (which might be ten to two hundred feet deep) was removed, the soft, earthy ore was dug out, at first by individual miners and later by massive steam-powered shovels. Techniques developed to mine the Mesabi were later used in open-pit copper mines in Arizona and Utah.

The town of Hibbing, Minnesota, became the principal commercial center of the Mesabi district, but numerous other mining communities came and went as railroads were built and mines dug. Some of these towns were company owned and controlled. Others were independent settlements. Most were populated largely by European immigrant workers who arrived to do the hard physical labor of producing the iron ore. Mining on the Mesabi was seasonal work, since the ground could only be worked between the spring thaw and the winter snows. Activity ceased during the long Minnesota winters, when the Great Lakes were frozen, preventing the ore from being shipped to the industrial centers where it was processed into steel.

Despite occasional financial crises, including the Panic of 1893 that forced the Merritt brothers into bankruptcy, commercial exploitation of the range progressed rapidly. Both John D. Rockefeller and Andrew Carnegie bought Mesabi properties, and large parts of the range were eventually incorporated by J. P. Morgan into the U.S. Steel Corporation, formed in 1901. Fortunes were being made or augmented by producing and shipping the iron, but the physical consequences for the landscape were equally striking. In the words of one history of the region, "Once such wealth was discovered, the face of the land changed irrevocably. In only a few years the magnificent virgin forests of red and white pine and spruce were gone, leaving miles of stump-littered, barren hills. Gaping red canyons were gouged out of the land, the lean ore dumps forming man-made mountains . . . Nowhere else in the nation have changes of such magnitude occurred as quickly." The remains of these pits and the accompanying ore dump hills characterize the landscape of the modern Mesabi. Their use and modification reveals landscape, as W. J. T. Mitchell has urged, "as a cultural practice, not merely symbolizing but embodying cultural power."[6]

The region's iron ore deposits seemed inexhaustible when first discovered, but a growing appetite for steel, combined with the defense needs of World War II, soon diminished the supply of what was known as "natural" ore. Productive mines gradually stripped away the last of this high-grade ore, and the final natural-ore mine on the Mesabi ceased operation during the 1990s. Iron mining did not altogether stop, however, since the range still contains significant deposits of a low-grade, hard-rock form of iron ore called taconite.[7]

Foreseeing the exhaustion of the natural ore, companies such as U.S. Steel began in the 1940s to subsidize experiments with taconite. The low percentage of iron contained in taconite had caused it to be scorned by earlier generations. It was more difficult to mine than the natural ore and couldn't be used in existing blast furnaces. After considerable experimentation, it was discovered that the ore could be used if it were first crushed and concentrated and then processed into pellet form. Mesabi companies began to invest in taconite mining and processing after a tax break from the state, in the form of a constitutional amendment guaranteeing favorable taxation of taconite. As a result, Minnesota is today the largest producer of iron ore and taconite pellets in the United States and also suffers the environmental consequences of the dumping of taconite tailings into the waters of Lake Superior.

By now, the Mesabi Range has been mined intensively for more than one hundred years. Geologists estimate that one hundred more years' worth of taconite ore still remain in reserve. Iron mining and processing thus continue to be important activities in the region, even as they have massively transformed the physical environment. Yet, as the state government of Minnesota cheerfully notes in its guide to minerals, "Mining is a *temporary* use of the land."[8] Thus the salient question for the Mesabi is, What kind of use *can* follow a pit mine? Just exactly how *does* one recycle a mining landscape?

Responses to these questions are embedded in the physical landscape, and in the various promotional and framing devices that surround it for public consumption. Tour guides and visitors' brochures now formulate the Mesabi landscape alternately as a technological wonderland or a recovering natural area. An industrial theme park, IronWorld Discovery Center, maps the Mesabi as a kind of historical food court, where immigrant cultures are manifested as distinctive cuisines. The photographs that follow are suggestive, not comprehensive, but they reflect the range of meanings attributed to the mining landscapes of northeastern Minnesota.

These meanings, in turn, are informed by differing assumptions about the cultural role of mining and about place. As Doreen Massey has pointed out, "The identities of places are always unfixed, contested and multiple." Where Mumford finds barbarism, longtime residents of the Mesabi take pride in a transformed landscape and the productive work that it represents. Both views have consequences for the landscape, as the work of environmental psychologists demonstrates. Environmental restoration privileges the natural and works to obscure traces of Mumford's social degradation. Such attitudes mandate disguise, or at least mitigation, of the physical legacy of mining. Industrial tourism, in contrast, celebrates the presence of humans. It enshrines the massive steam shovels and trucks that were the tools of transforming the landscape and builds viewing platforms for the abandoned pits. Manifested on the Mesabi, then, are a number of competing landscape identities, no one of which prevails.[9]

In Figure 14.2, a typical Mesabi mining landscape reveals the contoured sides of a mined pit, waste dumps made up of displaced overburden and unprocessed rock, and an abandoned pit filled with water. Because of the high water table in the area, mines had to be pumped constantly in order to be worked. Once the pumping stopped, they filled naturally with water. Some of these now serve as municipal water supplies or are stocked with fish for recreational use.

Figure 14.3 presents a panorama of reclaimed and "natural" mining sites near Hibbing, Minnesota, in 1996. Over time, revegetated areas begin to blend back into the hillocky contours of the surrounding landscape.

The former town site of North Hibbing, Minnesota, is shown in Figure 14.4. Hibbing was a mining boomtown founded in 1893 as a collection of tents, tarpaper shacks, and log cabins. By 1910, it was a bustling city of sixteen thousand people. As the demand for Mesabi iron increased, nearby mines expanded to surround Hibbing on three sides. Eventually, the value of the ore forced relocation of the town to a site two miles south, a move that was complete by 1950. The old town site was then called North Hibbing. It is now a city park, complete with curbs and stately trees, nestled incongruously amid the mines that created it.

Figure 14.5 shows the viewing platform for the Hull-Rust Mahoning Mine. Hibbing, meanwhile, promotes visitorship at the Hull Mahoning Mine complex, the largest open-pit iron mine in the world. In the words of the *Hibbing Visitors' Guide,* this mine, still operated by the Hibbing Taconite Company, "stands as a monument to the ingenuity—and just plain hard work—of man." The guide recommends the view from the observation complex, which shows "the vast pit more than 3 miles long, up to 2 miles wide, and 600 six hundred feet deep." The rhetoric of the guide and the tourism it promotes depend on qualities of gigantic size, scale, and ambition delineated by David Nye as part of the American technological sublime. From this perspective, the vestiges of a mining landscape are venerable artifacts of human and technological productivity. They are worthy of preservation for the very reason that they have had such tremendous impact on the landscape.[10]

Figure 14.6 depicts mining gigantism. Among the most common of the mining artifacts are the huge tools it requires. Pictured here are a 100-ton dump truck at the Rouchleau Group Mines near Virginia, Minnesota, and a Black Amsco steam shovel scoop in Mountain Iron, Minnesota, where the Merritt brothers made their 1890 discovery. Mountain Iron is now a National Historic Landmark. In both cases, the salient point of the technology is size.

Conveying scale in a mining landscape is difficult, so tools are almost universally used as a metaphor for the grandeur and complexity of the entire undertaking. Statistics on the sign in front of the truck wheel, for instance, carefully record the dimensions, cost, and weight of the truck and its tires. The tires are more than eight feet in diameter and cost $6,100 each. Note the presence of a similar huge tire behind the shovel scoop at Mountain Iron, sufficient even without accompanying statistics to convey an impression of enormity to the viewer. Here, the technological sublime is rendered palpable through the trope of the huge tire.

Figure 14.7 shows Judson Pit, Buhl, Minnesota. This reclaimed mine pit is maintained by the city of Buhl. Its status as a restored landscape is proudly proclaimed by the sign, which marks the site as a product of the Iron Range Resources and Rehabilitation Board (IRRRB). This public entity is funded by the mining industry and is charged with both reclamation and recycling of the unused mine dumps and pits that remain from natural-ore mining.

Minnesota takes pride in having been one of the first states in the United States to require comprehensive mineland reclamation for iron ore and taconite mines, in 1980. The Iron Mining Association of Minnesota extolls this reclamation work, pointing to the uses of abandoned pits as fishing lakes, municipal water supplies, wildlife refuges, and tourist attractions. Their brochure boldly claims that "mining is an asset to Minnesota's tourism industry," and they extoll their own stewardship of the Mesabi environment.

These claims rest on the assumption that the most appropriate Mesabi landscape is, in fact, a "natural" one, restored as closely as possible to its "original" appearance before the mines were dug. The Iron Mining Association's brochure promotes a vision of the modern Mesabi as virtually indistinguishable—at least eventually—from its original character as part of Minnesota's North Woods. In this vision, the economic history and social displacement represented by the North Hibbing and the Hull-Rust Mahoning Mine have no place, and the huge tires are discretely recycled instead of proudly displayed. Here, technology is employed to mask its own effects in the landscape, rather than to perpetuate its achievements.[11]

Modern mining tourism is depicted in Figure 14.8. On the top is an observation platform on the Minnesota Taconite Mine Company tour.

MinnTac is the largest taconite plant in the world, with an annual capacity of 18 million tons of taconite pellets. Visitors are given a tour of the modern mining operation, outfitted in orange helmets that clearly identify them as "pumpkin heads," or tourists. On the bottom is the Old Town Mine View in Virginia, Minnesota. Tourists avoid the need for helmets, because the caged-in lookout protects them from the potentially dangerous landscape. Here the visitors are overlooking a massive, heavily revegetated pit.

On the Mesabi, mining operates on numerous levels simultaneously: it is actively pursued in operating mines; it is commemorated in parks and roadside displays; through reclamation activities, it is erased visually from the landscape; and it is promoted as a tourist activity. The latter phase is depicted here, as mining landscapes are recycled, with the help of the IRRRB and the State of Minnesota, to become tourist destinations.[12]

Rendering mining appealing in the late twentieth century United States requires considerable effort and imagination. Where the landscapes are not natural but technological, tourists require substantial instruction to comprehend their import, to be persuaded that there is something to "see." Geographer Richard Francaviglia coined the term "technostalgia" to characterize tourists' fascination with the apparatus of nineteenth-century mining, especially in the American West, where head frames and mine dumps take on a vaguely picturesque, weathered quality in old photographs and movies. It is more difficult to find visual appeal in today's abandoned iron pits and waste dumps. Tour guides and framing devices help mark "views" that remain cryptic for the uninstructed visitor.[13]

Figure 14.9 shows the pump house and water truck and picnic table at the Godfrey-Glen Mine, IronWorld Discovery Center in Chisholm, Minnesota. This scene combines commemorative, educational, and recreational functions in a landscape explicitly designed to appeal to tourists. The sign explains the necessity for pumping at Mesabi mines and the role of the pump house and water truck in keeping a pit free of water. The picnic table implies a recreational potential otherwise not apparent at the site.

Part of the miniature golf course at IronWorld Discovery Center is shown in Figure 14.10. Run by the IRRRB, IronWorld is an industrial theme park founded in 1977. It features a working trolley imported from Australia, a Hall of Geology, ethnic food kiosks, mining equipment displays, remote-controlled ore boats, a homestead demonstration area, genealogical and historical displays, and interactive exhibits focusing on the history of mining. The miniature rock dump pictured here is designed to mimic the characteristics of the technological landscape beyond the gates of the theme park.

Ultimately, the Mesabi today incorporates a number of disparate visions of what a mining landscape is and means. Technology works simultaneously to create this landscape and to obliterate it in the interests of a more "natural" version. Residents and visitors encounter the same space within different perceptual contexts, so that its history may be vivid and immediate to one group but abstract and institutional to another. One group recreates in another's working space, as memory of ethnic tensions and labor unrest is effaced from the preserved landscape. As a case study, it confirms art historian Stephen Daniels's observation that "any apparently simple picture of a country scene may yield many fields of vision." On the Mesabi the apparently empty iron ore pits are, in fact, anything but.[14]

NOTES

1. Lewis Mumford, *Technics and Civilization* (1934; reprinted New York: Harcourt, Brace & World, 1963), 73.

2. Russell H. Bennett, *Quest for Ore* (Minneapolis, Minn.: T. S. Denison, 1963), 21.

3. The name *Mesabi* is reportedly an Ojibwa term meaning "sleeping giant." This account of the discovery and exploitation of the Mesabi Iron Range is drawn from Bennett, *Quest for Ore;* E. W. Davis, *Pioneering with Taconite* (St. Paul: Minnesota Historical Society, 1964); and Peter F. Torreano, *Mesabi Miracle: The 100-Year History of the Pillsbury-Bennett-Longyear Association* (Hibbing, Minn.: Sargent Land Company, 1991).

4. *Iron Range Country: A Historical Travelogue of Minnesota's Iron Ranges* (Eveleth, Minn.: Iron Range Resources and Rehabilitation Board, 1979), 99.

5. J. H. James quoted in *The Great Missabe Iron Range,* pamphlet, (Mines Library, University of Nevada, Reno, 1892).

6. *Iron Range Country,* 91–92; W. J. T. Mitchell, Introduction to *Landscape and Power,* ed. W. J. T. Mitchell (Chicago: University of Chicago Press, 1994), 2.

7. For descriptions of natural ore and taconite, see Davis, *Pioneering with Taconite.*

8. *Minnesota Minerals* (St. Paul: Minnesota Department of Mines and Minerals, 1994), 5.

9. Doreen Massey, *Space, Place, and Gender* (Minneapolis: University of Minnesota Press, 1994), 5. Massey points out that place is also inextricably bound up in time, so that it is the "unutterable mobility and contingency of space-time" that is being contested in places like the Mesabi. For the impact of attitudes on the physical environment, see Reginald G. Golledge, "Cognition of Physical and Built Environments," in *Environment, Cognition, and Action: An Integrated Approach,* ed. Tommy Garling and Gary W. Evans (New York: Oxford University Press, 1991), 35–62. Kent C. Ryden, in *Mapping the Invisible Landscape* (Iowa City: University of Iowa Press, 1992), discusses a similar sense of pride among residents in the Coeur d'Alene mining district of Idaho.

10. Hull Mahoning Mine tour pamphlet and *Hibbing Visitor's Guide,* both in authors' possession. David Nye, *American Technological Sublime* (Cambridge, Mass.: MIT Press, 1994). For Americans, the technological sublime was also linked with nationalism, as Nye explains: "The sublime was inseparable from a peculiar double action of the imagination by which the land was appropriated as a natural symbol of the nation while, at the same time, it was being transformed into a man-made landscape" (37). James Dickinson describes "a sort of late twentieth-century sublime" that awes us with "the speed with which change dissolves the past and constantly reshapes the once immutable landscape around us" (82). James Dickinson, "Entropic Zones: Buildings and Structures of the Contemporary City," *Capitalism, Nature, Socialism* (September 1996): 81–95.

11. Iron Mining Association of Minnesota brochure, in authors' possession. For a visual and philosophical rumination on the paradox of engineered natural landscapes, see Peter Goin, *Humanature* (Austin: University of Texas Press, 1996).

12. For an intriguing exploration of the process by which locations may acquire new identities and evolve from ruins into monuments, see Dickinson, "Entropic Zones."

13. Richard Francaviglia, *Hard Places* (Iowa City: University of Iowa Press, 1992). Authorities in Anaconda, Montana, have taken a similar approach, creating a golf course on a Superfund site that marks the former location of a copper smelter. According to the *New York Times,* the Old Works Golf Course is a recreational reclamation project with "an industrial smelting theme that carries right down to the black slag, or smelting waste, used in the bunkers and the cranes and old heavy equipment scattered around the grounds" (6 April 1997, 20).

14. Stephen Daniels, *Fields of Vision: Landscape Imagery and National Identity in England and the United States* (Princeton, N.J.: Princeton University Press, 1993), 8.

FIGURE 14.2 Peter Goin, copyright 1996. All rights reserved.

FIGURE 14.3 Peter Goin, copyright 1996. All rights reserved.

FIGURE 14.4 Peter Goin, copyright 1996.

FIGURE 14.5 Peter Goin, copyright 1996.

FIGURE 14.6 Peter Goin, copyright 1996. All rights reserved.

FIGURE 14.7 Peter Goin, copyright 1996. All rights reserved.

FIGURE 14.8 Peter Goin, copyright 1996.

FIGURE 14.9 Peter Goin, copyright 1996. All rights reserved.

FIGURE 14.10 Peter Goin, copyright 1996.

Notes on Contributors

David E. Nye (Ph.D. in American studies, University of Minnesota) is professor and chair at the Center for American Studies, Odense University, Denmark. His many books include *Consuming Power, Narratives and Spaces, American Technological Sublime,* and *Electrifying America,* which received the Dexter Prize. His current research concerns technology, landscape, and narrative.

Paul Brassley studied agricultural economics at the University of Newcastle and agricultural history at Oxford. He is senior lecturer in agricultural history and policy in the Seale-Hayne Faculty of the University of Plymouth. Author of *Agricultural Economics and the CAP: An Introduction,* his current research concerns rural history, culture, and landscape.

James Dickinson (Ph.D. in sociology, University of Toronto) teaches at Rider University. Recent publications include "Entropic Zones: Buildings and Structures of the Contemporary City," in *Capitalism, Nature, Socialism.* His current research focuses on art and technology, the future of industrial ruins and the recycling of cities.

Jacob Wamberg (Ph.D. in art history, Copenhagen University) is assistant professor at the Institute of the History of Art, Aarhus University, Denmark. His publications include articles on fifteenth-century landscape and the editing (with H. D. Christensen and A. Michelsen) of *Kunstteori, Positioner i nutidig kunstdebat* (in Danish, on contemporary art theory.)

TADEUSZ RACHWAL is associate professor of English, University of Silesia, Poland. The co-author of a book on Jacques Derrida in Polish, his English publications include *Word and Confinement: Subjectivity in 'Classical' Discourse* and *Approaches to Infinity: The Sublime and the Social.*

STUART KIDD, until recently director of American Studies at the University of Reading, is now Warden of Bulmershe Hall. The author of many essays on American cultural history during the interwar period, he recently co-edited *The Roosevelt Years: New Perspectives on the United States, 1933–45.*

CHRISTOPHER BAILEY (B.A. in art history, University of East Anglia) is professor and head of the Department of Historical and Critical Studies, University of Northumbria. The editorial secretary of *Journal of Design History,* he has published on rural industries, economic regeneration, and urban technological networks. His current research includes the role of contemporary crafts in the economy of north-eastern England.

STEPHEN MOSLEY (Ph.D. in history, University of Lancaster) wrote his dissertation on smoke pollution in Victorian and Edwardian Manchester. He is a member of the history department, University of Birmingham, Westhill, and is now writing a comparative study of air pollution in nineteenth-century Britain and Germany.

BARBARA ALLEN (architect, Columbia University; Ph.D. in Science and Technology Studies, RPI) holds the Contractors Educational Trust Fund Chair in Architecture at the University of Southwestern Louisiana. She is executive editor of *The Journal of Architectural Education.* Recent articles have appeared in *Radical History Review* and *Michigan Feminist Studies* and also include "The Popular Geography of Illness," in *Centuries of Change.*

MARK LUCCARELLI (Ph.D. in American studies, University of Iowa) is associate professor in the Department of British and American Studies at the University of Oslo. He is author of *Lewis Mumford and the Ecological Region.* His current research concerns regional planning, the idea of landscape and environmental reform.

Thomas Zeller (Ph.D. in history, Munich) is visiting assistant professor in the School of History, Technology and Society at Georgia Institute of Technology. He has also been a research associate in science and technology at the Deutsches Museum. His dissertation on the construction of roadways, railways, and landscapes in twentieth-century Germany will be published by Campus.

Stephen Bending (Ph.D. in English, University of Cambridge) is a lecturer at the University of Southampton. His most recent essays are on antiquarianism, the country house, and eighteenth-century gardens; he is co-editor of the forthcoming anthology *The Writing of Rural England, 1500–1800.*

Pia Maria Ahlbäck is a doctoral student and teacher at the department of English at Åbo Akademi University, Turku, Finland. She specializes in literature and the environment, with a focus on images of energy in the work of George Orwell. She also has worked as a journalist and is a prize-winning short story writer.

Peter Goin (M.F.A., University of Iowa) is professor of art in photography and video at the University of Nevada, Reno. His numerous books include the award-winning *Nuclear Landscapes* and *Humanature.* Recently, he collaborated with the Center of the American West in Boulder, Colorado, on the *Atlas of the New West.* He has exhibited internationally in more than fifty museums and received two National Endowment for the Arts Fellowships.

Elizabeth Raymond (Ph.D. in American civilization, University of Pennsylvania) chairs the Department of History at the University of Nevada, Reno. She co-edited, with Ronald M. James, *Comstock Women: The Making of a Mining Community.* She has collaborated with Peter Goin on *Stopping Time: A Rephotographic Survey of Lake Tahoe* and *Physical Graffiti: Mining in America,* forthcoming.

Index

aesthetics, 9–10, 13, 42–43, 62, 71–75, 93, 242, 247–248. *See also* entropy; landscape; sublime
African Americans, 189–201
agriculture, 6, 11, 15; art and, 74, 76–80; English, 21–28, 69–71, 138, 241–251; U.S., 99, 106–111, 121–123, 125, 189
Anderson, Sherwood, 127–128
Andrews, Thomas, 175
Appalachian Trail, 8, 121, 207–216
autobahn, 7, 8, 218–233
automobiles, 128, 129, 214, 220, 230

Bachelard, Gaston, 213
Bartram, William, 90–95
Baudrillard, Jean, 89
Bellasis, Mrs. E. J., 164–165
Bigelow, Jacob, 14
blight, industrial. *See* pollution
Bone, William, 180
Boorstin, Daniel, 87, 91
Bourdieu, Pierre, 75
Brooke, J. R., 144
Bureau of Reclamation, 107–109, 111
Byatt, A. S., 24

Campin, Robert, 71, 73
Carpenter, Edward, 174–175
Chaucer, 25
chemical corporations, 188, 190–198
Christos, the, 43, 48–54, 62
Clark, Kenneth, 40
Clarke, Allen, 180
Clements, Frederick, 228
Cohen, Anthony, 151
Collier, John, 122–123, 129–130
Colorado River, 97–113
Cosgrove, Denis, 31
Council for the Preservation of Rural England, 138–140, 151, 152
countryside: of American South, 119–133; English, 3, 8, 21–34, 136–154, 241–251; in painting, 69; in photography, 124–132
crafts, 142–145, 148–150
Cronon, William, 4, 162, 209, 210

dams, 106–113, 130–131, 211, 216
de Vries, Jan, 31
Delano, Jack, 120, 126, 129, 131
Denmark, 3, 15
Dennison, S. R., 138–139, 145
Department of Environmental Quality, 190, 192, 195–199
desert, 76, 98–99, 107
Design & Industries Association, 151–152
Dickens, Charles, 170

disease, 175–179, 181–182, 193
Drake, Daniel, 10–12
Duchamp, Marcel, 42
Deutsche Technik, 223–224, 230
Dutton, Clarence, 101

ecology, 188, 213–215, 218–219, 232–233. *See also* landscape; nature
enclosure, 244
engineers, 105, 108–109,
England, 136–201, 241–265
entropy, 55–57, 172
ephemerality, 6, 7, 21–22, 24–28, 31–34, 43–44, 46–47, 53, 56, 77. *See also* recycling
European Union, 26
Evans, Walker, 123, 126, 128
exploration, 87–95, 103–111, 211–212. *See also* tourism

factories, 123–124, 130, 163–164, 166, 169–170, 254–256, 264
Farm Security Administration, 8, 119–135
fields, 21–28, 31, 69, 71, 74, 79, 80, 91, 94, 99, 121–123, 125–126. *See also* landscape
Fish, Stanley, 188
forests, 4, 15, 28, 92, 211, 260. *See also* agriculture; wilderness
Foucault, Michel, 201, 255–256, 262

galleries, 41–43, 60–62
Garden of Eden, 70, 76, 248
gardens, 91, 94, 137, 178–179, 210, 241–242, 245, 248–251
Germany, 218–233
Gillette, King Champ, 112
Gilpin, William, 245
Goetzmann, William H., 103
Gombrich, Ernst, 74
Grand Canyon, 7–8, 55, 97, 101, 103, 105, 111

Haraway, Donna, 188, 200
Hart, Ernest, 179–180
Harvey, David, 33
Heimatschutz movement, 222–224
Heinz, 254–258, 264
Hennell, Thomas, 152–154
heterotopia, 7, 255–256, 262, 265n5
Hibbing, Minn., 269
Hillers, Jack, 104
Hitler, Adolf, 222
homesteading, 107–108
Hoover Dam, 98, 109–111, 113
hydroelectricity, 109–111, 211

imperialism, 87–96
industrialism, 49, 56, 123, 129–131, 136, 162–164, 190–201, 211, 254–264, 269–271
industries, rural. *See* Rural Industries Bureau
irrigation, 106–110
Ives, Joseph C., 98–99

Jackson, J. B., 3
Jussim, Estelle, 13

King, Clarence, 99–100
Kuhn, Thomas, 74, 75

Ladurie, Roy, 81
landscape: architecture, 224–232; art and, 41–63; "authentic," 11, 31, 55–60, 122, 123; biology and, 29–30; class and, 242–251; defined, 3–6, 9–10, 13–16, 31, 69–75, 190–191; dialectical, 54, 57; ephemeral, 21–34, 43–56, 77; history of, 22–24; industrial, 14, 123–124, 161–201, 254–264, 270; painting, 9, 69–82; photographic, 13, 119–129, 267–277; as prospect, 213–214; sexual, 91–94; time and, 77; toxic, 187–203; voids in, 55; work in, 5, 77–81. *See also* agriculture; forests; maps; nature; sublime
Lange, Dorothea, 124–126, 128
language, 87–95
Las Vegas, 110
Lee, Russell, 121, 126, 127
Leopold, Aldo, 213

Lévi-Strauss, Claude, 89
Linnaeus, 91–93, 95
livestock, 27–28
Louisiana, 189–201
Lowell, Mass., 12

MacKaye, Benton, 207–216
MacLeish, Archibald, 125
Manchester, Eng., 9, 161–201
Manchester Association for the Prevention of Smoke, 161, 170–171
Manifest Destiny, 97, 106, 110
maps, 79, 88–91, 101, 200
Marx, Leo, 14, 210
Massingham, H. J., 152–154
mass media, 128
mechanization, 124,
Merchant, Carolyn, 210, 211
Mesabi Iron Range, 267–275
Middle Ages, 70–71, 77–80
migration, western, 102, 110
Miller, Perry, 212
mines, 7, 90, 102–103, 104, 123, 263, 267–276
Mississippi River, 9, 187,
modernization, 126, 136, 137
Molesworth, John, 161, 174
Monaco, Lorenzo, 71–72
Moran, Thomas, 101
Mumford, Lewis, 210, 267, 271
Mydans, Carl, 120

Napier, Charles, 163
narrative, 162, 167–168, 170, 174–179, 183, 187–190, 198, 200
Native Americans, 16, 89–90, 95, 101–103, 106, 112
natural history, 91–95
nature: aesthetics and, 9–11, 69–75; authenticity and, 11, 31, 55–60, 122, 123, 225–228, 267–276; classifying, 90–95; domination of, 109–112, 267–271; engineering of, 207–234. *See also* landscape; maps; space
Nazi Germany, 218–233

Netherlands, The, 15–16
Neumann, Erich, 75
New Deal, 119–133
New England, 3, 15
Niagara Falls, 112
Nicholson, Marjorie, 10
Noble, John W., 110
Nye, David, 131, 212, 272

O'Sullivan, Timothy, 102
Office of War Information (U.S.), 130
Olwig, Kenneth, 210
Oppenheim, Dennis, 41, 43–48, 53, 62
Orwell, George, 12, 254–264
Ovid, 79

painting, 9, 34, 40, 69–82, 101–102, 120
parks, 112, 272–274. *See also* tourism
pastoralism, 10–11, 21–22, 211. *See also* landscape; nature
pathways, 207–235
perspective, linear, 100–101, 74–79, 111–112
photography, 102, 104, 119–133, 271–275
Picasso, Pablo, 71
pipelines, 130, 191
plant sociology, 228–229,
plantations, southern, 121–122
Plato, 78, 89
pollution, 3, 9, 55–57, 123–124, 129–130, 161–203, 259–261, 263–264, 269–275
Porter, Robert, 103
Post Wolcott, Marion, 123–124
Powell, John Wesley, 100, 103–104, 106, 108, 111
Pratt, Mary Louise, 92, 94
Pyne, Stephen J., 101

railroads, 105, 128, 225
Ransome, Arthur, 175–177
recycling, 271–275
Renaissance, 81
Rew, Sir Henry, 146
Rural Community Councils, 136–137, 145–148

Rural Industries Bureau, 136, 138, 140–150
Rural Life Conference (1920), 146–147

Schama, Simon, 4
Seifert, Alwin, 224–226, 228
Sharp, Thomas, 140
Smithson, Robert, 41, 43, 55–62
smoke, 161–183
South, American, 8, 119–135, 187–204
space: geometric, 111–112; imaginative, 120, 213; in painting, 71–75; as site and nonsite, 41–62. *See also* maps; painting; photography
Standard Oil, 130
Stanton, Robert, 105–106, 111
Steinbeck, John, 124
Straub, Marianne, 141
Stryker, Roy, 119–121, 123, 127–130, 132
sublime, 10, 56, 93, 100, 105–106, 111, 113, 131, 213, 272
Sutton, Ann and Myron, 207, 208

Taylor, Paul, 124–126
Taylor, William Cooke, 164
technology: definition of, 14–16, 87; landscape and, 10–16, 262–263, 271–275; threat of, 5, 215, 222–223, 262. *See also* agriculture; chemical corporations; dams; *Deutsche Technik;* fields; industry; language; maps; mines; photography; pipelines; railroads
Tennessee Valley Authority, 127, 130–131
textile manufacturing, 123–124, 163–164, 166, 169–170
Thompson, Flora, 151
Todt, Fritz, 222–227, 229
tourism, 7, 11–12, 102, 213–215, 221–241, 245–251, 271–275

U.S. Steel Corporation, 269, 270
utilitarianism, 106–113
utopia, 112, 262. *See also* heterotopia

Vachon, John, 130, 131, 132
Veblen, Thorstein, 14
villages, 127–128, 146–147
Vincent, E. R., 148–149, 157n32
volcanoes, 7

Waugh, Edwin, 167–168
Waters, Frank, 113
Weber, Max, 81
Wheeler, George M., 100, 102, 103, 106, 111
Wigan, Lancashire, 12, 254–264
wilderness, 9–10, 12, 122, 207–208, 209, 211, 213
Williams, W. M., 141, 148
Wolf, George, 127
work, 5, 23–24, 27, 69, 71, 74, 77–81, 111, 142–145, 148–149, 163–170, 181–182
Worster, Donald, 100, 107–108

Yellowstone National Park, 112
Young, Arthur, 241–251